Ion-selective electrodes

SECOND EDITION

JIŘÍ KORYTA

*Professor of Physical Chemistry, Charles University, Prague,
and Head of Membrane Electrochemistry Group,
J. Heyrovský Institute of Physical Chemistry and Electrochemistry,
Czechoslovak Academy of Sciences*

KAREL ŠTULÍK

Lecturer in Analytical Chemistry, Charles University, Prague

CAMBRIDGE UNIVERSITY PRESS

Cambridge

London   New York   New Rochelle

Melbourne   Sydney

CAMBRIDGE UNIVERSITY PRESS
Cambridge, New York, Melbourne, Madrid, Cape Town, Singapore, São Paulo, Delhi

Cambridge University Press
The Edinburgh Building, Cambridge CB2 8RU, UK

Published in the United States of America by Cambridge University Press, New York

www.cambridge.org
Information on this title: www.cambridge.org/9780521110761

First published 1975
Second edition 1983
This digitally printed version 2009

A catalogue record for this publication is available from the British Library

Library of Congress Catalogue Card Number: 82-25297

ISBN 978-0-521-23873-1 hardback
ISBN 978-0-521-11076-1 paperback

CONTENTS

Contents

PREFACE

The second edition of *Ion-Selective Electrodes* contains a survey of the theory
and applications of ion-selective electrodes based on the literature published up
to mid-1981. Because of the rapid progress in the whole field and the very large
amount of diverse data, a compact and unified treatment of the theory has been
attempted. The technology and the applications have also been updated. In view
of these facts we have had to write practically a new book. In contrast to the first
edition, only a selective list of references could be included in the book, because
otherwise we would have to deal with more than four thousand references.

We are grateful to our colleagues Professor J. Dvořák (Faculty of Science,
Charles University, Prague), Dr J. Bureš (The Institute of Physiology, Czechoslovak
Academy of Sciences, Prague), Dr J. Veselý (The Geological Survey, Prague) and
to Dr Z. Samec (The J. Heyrovský Institute of Physical Chemistry and Electrochemistry, Czechoslovak Academy of Sciences, Prague) for valuable suggestions
during the preparation of both editions of the book. The Czech manuscript
has been translated into English by Dr M. Hyman-Štulíková. We would also like
to thank Dr A. Kejharová, Mrs M. Kozlová, Mrs D. Tůmová and Mrs L. Korytová
for their help with the preparation of the manuscript.

We wish to express our gratitude to the following authors and organizations
for kindly permitting us to reproduce figures from their publications: Professor
J. N. Butler, Dr W. L. Duax, Dr B. Fleet, Professor G. G. Guilbault, Professor
P. Läuger, Dr G. Mattock, Dr L. Pioda, Professor G. Rechnitz, Professor W. Simon,
Professor K. Sollner, Professor M. Truter, Professor D. W. Urry and Professor
R. J. P. Williams, The American Association for the Advancement of Science,
the American Chemical Society, M. Dekker, Inc., New York, The Orion Research
Corporation, Inc., Cambridge, Mass, and the Society for Analytical Chemistry.

Jiří Koryta
Karel Štulík

Prague, May 1982

1

INTRODUCTION

There is a tendency in many fields of contemporary physical and chemical science to seek impulses from biology for solving problems. This tendency shows promise, not only because it leads to new artificial tools, materials or processes, but also because it forms models permitting deeper understanding, at least in certain areas, of the actual processes occurring in nature. Efforts to imitate nature thus facilitate its deeper understanding.

Ion-selective electrodes are a remarkable product of this approach. Their development can be followed in several circular pathways from natural bio-electric phenomena to artificial membrane systems and back again, to attempts to explain processes at a cellular level.

The relationship between electric and physiological processes was discovered in 1791 by L. Galvani [27] in his classic experiments with frog muscles and nerves. In spite of these exciting results, the study of electrophysiological phenomena did not progress for several decades. M. Faraday [99] studied the electricity produced by the torpedo, but only in order to prove that this 'animal' electricity is the same as other kinds of electricity, for example 'voltaic' electricity from an electric machine or 'galvanic' electricity from a galvanic cell.

In 1848 du Bois-Reymond [21] suggested that the surfaces of biological formations have a property similar to the electrode of a galvanic cell and that this is the source of bioelectric phenomena observed in damaged tissues. The properties of biological membranes could not, however, be explained before at least the basic electrochemistry of simple models was formulated. The thermodynamic relationships for membrane equilibria were derived by Gibbs in 1875 [29], but because the theory of electrolyte solutions was formulated first by Arrhenius as late as 1887, Gibbs does not mention either ions or electric potentials.

A further important step forward was the work of Nernst [73, 74] and Planck [81, 82] on transport in electrolyte solutions. Here the concept of the diffusion potential was defined; diffusion potential arises when the mobilities of the electrically-charged components of the electrolyte are different and is important both for description of conditions within membranes as well as for quantitative determination of the liquid–junction potential.

A basic property of the membrane surface, the potential at the membrane/ solution boundary, was defined and determined by Nernst and Riesenfeld [75, 85, 86] for the phase boundary between two immiscible liquids.

Membrane electrochemistry proper began in 1890 with the work of W. Ostwald [79], who formulated the concept of a semipermeable membrane that selectively affects the membrane transport of ions. According to Ostwald, 'not only electric currents in muscles and nerves, but also the mysterious phenomenon of electric fishes will finally find an explanation in the properties of semipermeable membranes'.

At the same time (1902), a membrane theory was proposed for the electrical properties of cells and tissues by Overton [80] and Bernstein [8], whose principles remain valid to the present day.

At the turn of the century, considerable attempts were being made to find suitable membrane models. These models fall into two groups: compact, usually liquid ('oil') and solid membranes [10, 33, 62, 75]; and porous membranes [9]. At the very beginning of the study of compact membranes, the glass electrode was discovered [18, 34], whose membrane represented the first observation of marked selectivity for a particular type of ion, here the hydrogen ion. It is interesting that this first ion-selective electrode remains the best and most widely used of all such electrodes.

Although the problem of the liquid membrane potential was solved in principle by Nernst, a discussion developed in the ensuing two decades between Bauer [6], who developed the adsorption theory of membrane potentials, and Beutner [10, 11, 12], who based his theories on Nernst's work. This problem was finaly solved by Bonhoeffer, Kahlweit and Strehlow [13], and by Karpfen and Randles [49]. The latter authors also introduced the concept of the distribution potential.

The search for models of biological membranes among porous membranes continued in the twenties and thirties. Here, Michaelis [67] and Sollner (for a summary of his work, see [90]; for development in the field, [89]) should be mentioned. The existence and characteristics of Donnan membrane equilibria could be confirmed using this type of membrane [20]. The theory of porous membranes with fixed charges of a certain sign was developed by Teorell [93], and Meyer and Sievers [65].

On the basis of the older results of Gorter and Grendel [31], Danielli and Davson [19] developed a theory in 1935 according to which the membranes of cells and cellular organelles are very thin formations composed of a bimolecular phospholipid layer covered on both sides by adsorbed protein. Further progress was made with the theory of membrane potentials at excitable (especially nerve) cells, developed by Goldman [30] and Hodgkin, Huxley and Katz [35, 36, 50]. Bilayer lipid membranes obtained first by Mueller and coworkers [72] are important models of biological membranes.

Research on thicker membranes has progressed in two directions since the thirties. It was apparent that the theory of the glass electrode based on the concept of the concentration cell or membrane potential with a diffusible hydrogen ion cannot explain certain of its properties, notably its marked selectivity for sodium ions, which is very strong for glasses of certain composition [60]. This fact led Nikolsky and Tolmacheva [77] to develop the ion-exchange theory of the glass electrode. The formation of a diffusion potential within the glass membrane was included in a perfected theory of the glass electrode by Eisenman and coworkers [22]. The new construction principle is included in gas probes with a hydrophobic membrane permeable for gases and with a glass internal electrode [88].

After many attempts to obtain a membrane electrode based on various kinds of crystalline materials, Pungor and Hallós-Rokosinyi [83] were finally successful in preparing the first practical ion-selective electrode with a heterogeneous membrane containing silver halide precipitate (Mirnik and Težak [68] made important earlier discoveries in this area). A real success in this field was the development of the first ion-exchange membrane from a single crystal by Frant and Ross [25]. Except for the glass electrode, their lanthanum trifluoride ion-selective electrode for determining fluoride is the most important sensor in this field.

Liquid membranes with dissolved ion-exchangers were used first by Sollner and Shean [91], but the first actual ion-selective electrode was constructed on this principle much later by Ross [87].

A different direction in ion-selective electrode research is based on experiments with antibiotics that uncouple oxidative phosphorylation in mitochondria [59]. These substances act as ion carriers (ionophores) and produce ion-specific potentials at bilayer lipid membranes [72]. This function led Štefanac and Simon to obtain a new type of ion-selective electrode for alkali metal ions [92] and is also important in supporting the chemi-osmotic theory of oxidative phosphorylation [69]. The range of ionophores, in view of their selectivity for other ions, was broadened by new synthetic substances [1, 61].

The ion-selective field-effect transistor (ISFET) represents a remarkable new construction principle [7, 63]. 'Inverse potentiometry' with ion-selective electrodes is the electrolysis at the interface between two immiscible electrolyte solutions (ITIES) [28, 55].

Biological principles are also used in enzyme electrodes, where the sensor (usually an ion-selective electrode) is covered by a polymeric carrier containing an enzyme [32]. The determinand reacts in the enzyme layer yielding a product that causes a signal in the sensor. The bacterium electrode is based on a similar principle [84], as are electrodes using tissue in place of the enzyme layer [2].

Research on ion-selective electrodes is progressing very rapidly and the literature contains more than four thousand publications (a practically complete survey of the literature up to the spring of 1982 can be found in four reviews in *Analytica Chimica Acta* [51-54]). Regular surveys of ion-selective electrode research are also given in *Analytical Chemistry* [13a-d, 25a, 96]. A number of books have been written on ion-selective electrodes [3-5, 14-17, 26, 38-40, 45, 46, 56, 57, 58, 64, 66, 70, 71, 76, 95, 97, 98]. A number of specialized symposia have been devoted to ion-selective electrodes and proceedings from them are often available in book form [17a, 23, 24, 37, 41-4, 47, 48, 94].

In the years from 1969 to 1972, Orion Research published a magazine devoted to developments in ion-selective electrodes [78], and, later, a guide to their products [34a]. Since 1979, the journal *Ion-Selective Electrode Reviews* has been published semi-annually [39a].

A great number of companies throughout the world manufacture ion-selective electrodes. In view of the rapid development of the field no survey of these companies will be presented in the present book.

References for Chapter 1

1 D. Ammann, E. Pretsch and W. Simon, *Anal. Lett.* **5**, 843 (1972).
2 M. A. Arnold and G. A. Rechnitz, *Anal. Chem.* **52**, 1170 (1980).
3 P. L. Bailey, *Analysis with Ion-Selective Electrodes*, Heyden, London (1976).
4 P. L. Bailey, *Analysis with Ion-Selective Electrodes*, 2nd edition, Heyden, London (1980).
5 G. E. Baiulescu and V. V. Cosofret, *Applications of Ion-Selective Membrane Electrodes in Organic Analysis*, John Wiley and Sons, Chichester (1977).
6 E. Bauer and S. Kronman, *Z. Physik. Chem.* **92**, 819 (1917).
7 P. Bergveld, *IEEE Trans.Biomed. Eng.* **17**, 70 (1970).
8 J. Bernstein, *Pflüger's Arch. Ges. Physiol.* **92**, 521 (1902).
9 A. Bethe and T. Toropoff, *Z. Physik. Chem.* **88**, 686 (1914).
10 R. Beutner, *Z. Elektrochemie* **19**, 319 (1913).
11 R. Beutner, *Z. Elektrochemie* **19**, 467 (1913).
12 R. Beutner, *Z. Elektrochemie* **24**, 94 (1918).
13 K. F. Bonhoeffer, M. Kahlweit and H. Strehlow, *Z. Elektrochemie* **57**, 614 (1953).
13a R. P. Buck, *Anal. Chem.* **44**, 270R (1972).

41 *Ion-Selective Electrodes* (ed. E. Pungor and I. Buzás), Symposium held at Mátra-füred 1972, Akadémiai Kiadó, (1973).
42 *Ion-Selective Electrodes* (ed. E. Pungor and I. Buzás), 2nd Symposium held at Mátra-füred, 1976, Akadémiai Kiadó, Budapest (1977).
43 *Ion-Selective Electrodes* (ed. E. Pungor and I. Buzás), International Conference held in Budapest 1977, Akadémiai Kiadó, Budapest and Elsevier, Amsterdam (1978).
44 *Ion-Selective Electrodes* (ed. E. Pungor and I. Buzás), 3rd Symposium held at Mátra-füred 1980, Akadémiai Kiadó, Budapest (1981).
45 *Ion-Selective Electrodes in Analytical Chemistry*, Vol. 1 (ed. H. Freiser), Plenum Press, New York (1978).
46 *Ion-Selective Electrodes in Analytical Chemistry*, Vol. 2 (ed. H. Freiser), Plenum Press, New York (1980).
47 *Ion-Selective Microelectrodes* (ed. H. J. Berman and N. C. Hebert), Plenum Press, New York (1974).
48 *Ion-Selective Microelectrodes and Their Use in Excitable Tissues* (ed. E. Syková, P. Hník and L. Vyklický), Plenum Press, New York (1981).
49 F. M. Karpfen and J. E. B. Randles, *Trans. Faraday Soc.* **49**, 823 (1953).
50 B. Katz, *Nerve, Muscle and Synapse*, McGraw Hill, New York (1966).
51 J. Koryta, *Anal. Chim. Acta* **61**, 329 (1972).
52 J. Koryta, *Anal. Chem. Acta* **91**, 1 (1977).
53 J. Koryta, *Anal. Chim. Acta* **111**, 1 (1979).
54 J. Koryta, *Anal. Chim. Acta* **139**, 1 (1982).
55 J. Koryta, *Electrochim. Acta* **24**, 293 (1979).
56 J. Koryta, *Ion-Selective Electrodes*, Cambridge University Press, Cambridge (1975).
57 J. Koryta, *Ions, Electrodes and Membranes*, John Wiley & Sons, Chichester (1982).
58 N. Lakshminarayanaiah, *Membrane Electrodes*, Academic Press, London (1976).
59 H. A. Lardy, D. Johnson and W. C. McMurray, *Arch. Biochem. Biophys.* **78**, 587 (1958).
60 B. Lengyel and E. Blum, *Trans. Faraday Soc.* **30**, 461 (1934).
61 R. J. Levins, *Anal. Chem.* **43**, 1045 (1971).
62 R. Luther, *Z. Physik. Chem.* **19**, 529 (1896).
63 T. Matsuo and K. D. Wise, *IEEE Trans.Biomed. Eng.* **21**, 485 (1974).
64 *Medical and Biological Applications of Electrochemical Devices* (ed. J. Koryta), Wiley, Chichester (1980).
65 K. H. Meyer and J. F. Sievers, *Trans. Faraday Soc.* **33**, 1073 (1937).
66 D. Midgley and K. Torrance, *Potentiometric Water Analysis*, J. Wiley and Sons, Chichester (1978).
67 L. Michaelis, *Kolloid-Z.* **62**, 2 (1933).
68 M. Mirnik and B. Težak, *Arkiv kem.* **23**, 1 (1951).
69 P. Mitchell, *Chemiosmotic Coupling and Energy Transduction*, Glynn Research, Bodmin (1968).
70 G. J. Moody and J. D. R. Thomas, *Selective Ion-Sensitive Electrodes. Selected Annual Reviews of the Analytical Sciences*, Vol. 3 (ed. L. S. Bark), The Chemical Society, London (1973).
71 W. E. Morf, *The Principles of Ion-Selective Electrodes and of Membrane Transport*, Akadémiai Kiadó, Budapest and Elsevier, Amsterdam (1981).
72 P. Mueller, D. O. Rudin, H. T. Tien and W. C. Wescott, *Nature* **194**, 979 (1962).
73 W. Nernst, *Z. Physik. Chem.* **2**, 613 (1888).
74 W. Nernst, *Z. Physik. Chem.* **4**, 129 (1889).

13*b* R. P. Buck, *Anal. Chem.* **46**, 28R (1974).

13*c* R. P. Buck, *Anal. Chem.* **48**, 23R (1976).

13*d* R. P. Buck, *Anal. Chem.* **50**, 17R (1978).

14 K. Cammann, *Das Arbeiten mit Ionenselektiven Elektroden*, Springe Berlin (1973).

15 K. Cammann, *Das Arbeiten mit Ionenselektiven Elektroden*, 2nd ed Springer-Verlag, Berlin (1977).

16 K. Cammann, *Working with Ion-Selective Electrodes*, Springer-Verl (1979).

17 K. Cammann, *Zastosowanie Elektrod Jonoselektywnych* (Use of io electrodes), Wydawnictva naukovo – techniczne, Warsaw (1977)

17*a* *Chemically Sensitive Electronic Devices* (ed. J. Zemel and P. Bergv Sequoia, Lausanne (1981).

18 M. Cremer, *Z. Biol.* **47**, 562 (1906).

19 J. F. Danielli and H. Davson, *J. Cellular Comp. Physiol.* **5**, 495 (19

20 F. G. Donnan, *Z. Elektrochemie* **17**, 572 (1911).

21 E. Du Bois-Reymond, *Untersuchungen Über Thierische Elektrizitä* Berlin (1848); Vol. II/1, Berlin (1849); Vol. II/2, Berlin (1884)

22 G. Eisenman, D. O. Rudin and J. U. Casby, *Science* **126**, 871 (195

23 Elektrody Jonoselektywne, *Extended Abstracts of a Symposium (Analytical Chemistry Commission of the Polish Academy of S* M. Trojanowicz), Warsaw (1975).

24 Elektrody Jonoselektywne, *Extended Abstracts of the 2nd Symp(by the Analytical Chemistry Commission of the Polish Acaden* (ed. M. Trojanowicz), Warsaw (1979).

25 M. S. Frant and J. W. Ross, *Science* **154**, 1553 (1966).

25*a* G. H. Fricke, *Anal. Chem.* **52**, 259R (1980).

26 C. Fuchs, *Ionenselektive Elektroden in der Medizin*, Thieme, Stu†

27 L. Galvani, *De Viribus Electricitatis in Motu Musculari Comment*

28 C. Gavach and F. Henry, *J. Electroanal. Chem.* **54**, 361 (1974).

29 J. W. Gibbs, *Collected Works*, Vol. 1, p. 83, Longmans, Green an (1928).

30 D. E. Goldman, *J. Gen. Physiol.* **27**, 37 (1943).

31 E. Gorter and F. Grendel, *J. Exp. Med.* **41**, 439 (1925).

32 G. G. Guilbault, *Pure Appl. Chem.* **25**, 727 (1971).

33 F. Haber, *Ann. Phys.* (4), **26**, 927 (1908).

34*a* *Handbook of Electrode Technology*, Orion Research, Cambridg(

34 F. Haber and Z. Klemensiewicz, *Z. Physik. Chem.* **67**, 385 (190<

35 A. L. Hodgkin and A. F. Huxley, *J. Physiol.* London **117**, 500 (

36 A. L. Hodgkin and B. Katz, *J. Physiol.* London **108**, 37 (1949).

37 Ion and Enzyme Electrodes in Biology and Medicine, *Proc. Inte* 1975 (ed. M. Kessler, L. C. Clark, D. W. Lübbers, I. A. Silve† Urban and Schwarzenberg, München (1976).

38 *Ion-Selective Electrode Methodology*, Vol. 1 (ed. A. K. Covingt Boca Raton (1979).

39 *Ion-Selective Electrode Methodology*, Vol. 2 (ed. A. K. Covingt Boca Raton (1979).

39*a* *Ion-Selective Electrode Reviews* (ed. J. D. R. Thomas), **1**, (197 3, (1981), 4 (1982), Pergamon Press, Oxford.

40 *Ion-Selective Electrodes* (ed. R. A. Durst), National Bureau of Washington (1969).

75 W. Nernst and E. H. Riesenfeld, *Ann. Phys.* 9, 8, 616 (1902).
76 B. P. Nikolsky and E. A. Materova, *Ion-Selektivnyie Elektrody*, Khimiya, Leningrad (1980).
77 B. P. Nikolsky and T. A. Tolmacheva, *Zh. Fiz. Khim.* 10, 504 (1937).
78 *Orion Research Newsletter, Specific Ion Electrode Methodology*, Vol. 1 to 4.
79 W. Ostwald, *Z. Physik. Chem.* 6, 71 (1890).
80 E. Overton, *Plfüger's Arch. Ges. Physiol.* 92, 346 (1902).
81 M. Planck, *Ann. Phys. Chem.* N. F. 39, 196 (1890).
82 M. Planck, *Ann. Phys. Chem.* N. F. 40, 561 (1890).
83 E. Pungor and E. Hallós-Rokosinyi, *Acta Chim. Hung.* 27, 63 (1961).
84 G. A. Rechnitz, T. L. Riechel, R. K. Kobos and M. E. Meyerhoff, *Science* 199, 440, (1978).
85 E. H. Riesenfeld, *Ann. Phys.* 8, 609 (1902).
86 E. H. Riesenfeld, *Ann. Phys.* 8, 616 (1902).
87 J. W. Ross, *Science* 156, 1378 (1967).
88 W. Severinghaus and A. F. Bradley, *J. Appl. Physiol.* 13, 515 (1958).
89 K. Sollner, *The Early Developments of the Electrochemistry of Polymer Membranes, Charged Gels and Membranes* (ed. E. Selegny), Vol. I, D. Reidel, Dordrecht (1976).
90 K. Sollner, *J. Macromol. Sci.-Chem.* A3, 1 (1969).
91 K. Sollner and G. M. Shean, *J. Am. Chem. Soc.* 86, 1901 (1964).
92 Z. Štefanac and W. Simon, *Chimia* 20, 436 (1966).
93 T. Teorell, *Trans. Faraday Soc.* 33, 1053 (1937).
94 *Theory, Design and Biomedical Applications of Solid State Chemical Sensors* (eds. P. W. Cheung, D. G. Fleming, M. R. Neuman and W. H. Ko), CRC Press, West Palm Beach (1978).
95 R. C. Thomas, *Ion-Sensitive Intercellular Microelectrodes, How to Make and Use Them*, Academic Press, London (1978).
96 E. C. Toren Jr. and R. P. Buck, *Anal. Chem.* 42, 284R (1970).
97 J. Veselý, D. Weiss and K. Štulík, *Analysis with Ion-Selective Electrodes*, Ellis Horwood, Chichester (1978).
98 J. Veselý, D. Weiss and K. Štulík, *Analýza Iontově-Selektivnimi Elektrodami*, SNTL, Prague (1978).
99 L. P. Williams, *Michael Faraday*, Chapman & Hall, London (1965).

2

THE THEORY OF MEMBRANE POTENTIALS

In physics, an elastic two-dimensional plate is termed a membrane (Latin *membrana* = parchment) but in chemistry the term denotes a body, usually thin, which serves as a phase separating two other bulk phases. If this body is permeable to the same degree for all components of the adjacent phases and does not affect their mobility, then its only function is to prevent rapid mixing of the two phases. This is then termed a diaphragm. A real membrane must exhibit a certain selectivity, based on different permeability for the components of the two phases, and is then termed a semipermeable membrane. Membranes separating two electrolytes that are not permeable to the same degree for all ions are called electrochemical membranes. It is with these that we are concerned here.

Ion-selective electrodes are systems containing a membrane consisting basically either of a layer of solid electrolyte or of an electrolyte solution whose solvent is immiscible with water. The membrane is in contact with an aqueous electrolyte solution on both sides (or sometimes only on one). The ion-selective electrode frequently contains an internal reference electrode, sometimes only a metallic contact, or, for an ion-selective field-effect transistor (ISFET), an insulating and a semiconducting layer. In order to understand what takes place at the boundary between the membrane and the other phases with which it is in contact, various types of electric potential or of potential difference formed in these membrane systems must first be defined.

2.1   General relationships

Electric potential differences are formed at the boundary between two phases, at least one of which contains electrically-charged species (ions, electrons, or even dipoles). These differences are of two basic types. The inner potential of a given phase α, $\phi(\alpha)$, is the electrical work required for transfer of a unit positive charge (the charge of one mole of electrons multiplied by −1 or the charge of a mole of protons) from a point in a vacuum infinitely distant from the given phase into the given phase. The difference in the inner potentials of two phases α and β,

$$\Delta_\beta^\alpha \phi = \phi(\alpha) - \phi(\beta),$$

is termed the Galvani potential difference. Strictly, this quantity can be measured only when both phases have the same chemical composition, for example two wires of the same metal or two electrolyte solutions in the same solvent and with the same composition; however, extrathermodynamic procedures can sometimes be found to measure the Galvani potential between two chemically different phases.

The electrostatic or outer potential of a given phase $\psi(\alpha)$ is the electrical work required for transfer of a unit charge from a point in a vacuum infinitely distant from the given phase into the vicinity of that phase. The difference between the outer potentials of two phases α and β,

$$\Delta_\beta^\alpha \psi = \psi(\alpha) - \psi(\beta),$$

is termed the Volta potential difference. This quantity can usually be measured directly.

Galvani potential differences are usually the basis of potentiometric measurements. A simple example of a Galvani potential difference is the electromotive force (EMF) of a galvanic cell. This quantity is given by the Galvani potential difference between chemically identical metallic leads to the two electrodes of the cell. An example of a reversible galvanic cell is the system consisting of a hydrogen electrode and a silver chloride electrode in an aqueous solution of hydrochloric acid.

Cu	Pt	H_2	HCl, H_2O	AgCl	Ag	Cu	
1	2	3	4	5	6	7	(2.1.1)

According to the IUPAC convention, the individual phase boundaries are designated in the scheme by vertical lines. The EMF is given by the difference between the internal electrical potential of the right-hand metallic lead and the internal electrical potential of the left-hand metallic lead in the scheme (2.1.1), i.e.

$$E = \phi(7) - \phi(1). \tag{2.1.2}$$

The EMF, E, of a reversible galvanic cell is related to the change in the Gibbs energy ΔG in the cell reaction

$$E = -zF\Delta G, \tag{2.1.3}$$

where F is the Faraday constant and z is the charge number of the cell reaction.

This number gives the number of charges corresponding to a Faraday constant, i.e. the charge of 1 mole of protons that must pass through the cell in order for the cell reaction (the chemical reaction occurring in the cell on passage of an electric current) to occur to unit degree. The cell reaction corresponding to cell (2.1.1),

$$2AgCl + H_2 = 2Ag + 2HCl, \tag{2.1.4}$$

thus has a charge number of $z = 2$.

The electrode potential is the EMF of a cell in which the electrode on the right-hand side in the scheme is the test electrode, while the electrode placed on the left-hand side is a standard hydrogen electrode (i.e. a hydrogen electrode saturated with hydrogen under standard pressure $p_0 = 1.013 \times 10^5$ Pa and placed in a solution with an activity of hydronium ions equal to one).

In the field of ion-selective electrodes the potentials of electrodes of the second kind, acting as reference electrodes, are especially important. The electrode potentials of the most important reference electrodes, the silver chloride and calomel electrodes, are

$$E_{MCl,M} = E^0_{MCl,M} + (RT/F) \ln a_{Cl^-}, \tag{2.1.5}$$

where $E^0_{MCl,M}$ is the standard potential of the particular electrode of the second kind and a_{Cl^-} is the chloride ion activity.

So far we have considered electrodes whose potentials are determined through the cell reaction of the ions with which they are in contact. Such a potential cannot be formed on an ideally polarized electrode, for example a mercury electrode in a KCl solution within a certain potential region. In this case the electrode potential is determined by the electrode charge.

A cell whose equilibrium potential difference is affected by outer potentials is depicted in the scheme

$$\begin{array}{c|c|c|c|c}
\text{Ag} & \text{vacuum} & \text{KCl, H}_2\text{O} & \text{AgCl} & \text{Ag} \\
1 & 2 & 3 & 4 & 5
\end{array} \tag{2.1.6}$$

Apparently no cell reaction can occur in this cell as the vacuum prevents movement of species between phases 1 and 3. The measured difference in the inner electric potentials of phases 5 and 1 is equal to the difference in outer electric potentials between phases 5 and 3,

$$U = \phi(5) - \phi(1) = \psi(5) - \psi(3). \tag{2.1.7}$$

Detailed consideration of similar cases will be given in section 4.3 on ISFETs.

The electrochemical systems described so far are equilibrium systems. However, many membrane systems are not in equilibrium. Irreversible processes connected with charge transfer take part in the formation of electrical potential differences. If this charge transfer occurs in a homogeneous phase (for example, within the membrane), then the different velocities of the movement of the different kinds of ions leads to formation of a diffusion potential. An important type of diffusion potential is the potential difference formed at the contact between two electrolyte solutions with different compositions in the same solvent, called the liquid–junction potential. The formation of a subsequent potential difference at the membrane/electrolyte solution boundary may be influenced by slow ion exchange between the two phases.

The difference in inner electrical potentials between two electrolyte solutions separated by a membrane is termed the membrane potential, $\Delta\phi_M$. In the scheme

$$\text{solution 1} \quad | \quad \text{membrane} \quad | \quad \text{solution 2}$$
$$\phi(1) \quad x = p \qquad\qquad x = q \quad \phi(2) \qquad , \qquad (2.1.8)$$

where x is the coordinate perpendicular to the membrane and $\phi(1)$ and $\phi(2)$ are the internal potentials of solutions 1 and 2, then

$$\begin{aligned}\Delta\phi_M &= \phi(2) - \phi(1) \\ &= [\phi(2) - \phi(q)] + [\phi(q) - \phi(p)] + [\phi(p) - \phi(1)] \\ &= \Delta_q^2\phi + \Delta\phi_L + \Delta_1^p\phi, \end{aligned} \qquad (2.1.9)$$

where $\phi(p)$ and $\phi(q)$ are potentials inside the membrane close to phase boundaries with the solutions 1 and 2. In general these quantities are not equal, as will be shown in chapter 3, their difference $\phi(q) - \phi(p) = \Delta\phi_L$ is identical to the diffusion potential inside the membrane, see, for example, [3], or chapter 3 of [18].

2.2   The Nernst and distribution potentials

A particular component of a given phase can be characterized in terms of its content and ability to partake in various processes (chemical reactions, transport processes) using the partial molar Gibbs energy. For an electrically-charged phase, this quantity is termed the electrochemical potential of the ith component

$$\bar\mu_i(\alpha) = \mu_i^0 + RT \ln a_i + z_i F\phi(\alpha). \qquad (2.2.1)$$

Consider two phases, α and β, in contact, with a common ith component. Then

$$\bar\mu_i(\alpha) = \bar\mu_i(\beta), \qquad (2.2.2)$$

that is,

$$\mu_i^0(\alpha) + RT \ln a_i(\alpha) + z_i F\phi(\alpha) = \mu_i^0(\beta) + RT \ln a_i(\beta) + z_i F\phi(\beta). \qquad (2.2.3)$$

The relationship for the Galvani potential difference between the two phases follows from (2.2.3),

$$\Delta_\beta^\alpha\phi = [\mu_i^0(\beta) - \mu_i^0(\alpha)]/z_i F + (RT/z_i F) \ln [a_i(\beta)/a_i(\alpha)], \qquad (2.2.4)$$

where μ_i^0 is the standard chemical potential, z_i is the charge number and a_i is the activity of the ith component of the system. Quantity $\Delta_\beta^\alpha \phi$ is then termed the Nernst potential [1, 28, 33]. This type of potential difference cannot be obtained by purely thermodynamic methods (i.e. by measuring equilibrium potentials or by thermodynamic calculations at known compositions of both phases). Its calculation is prevented by the fact that the difference in the standard chemical potentials of the ith ionic component between phases α and β is not known. This difference is identical with the difference in the standard Gibbs energies of solvation between phases α and β, and is termed the standard Gibbs energy for transfer of the ith component from phase β to phase α,

$$\Delta G_{tr,i}^{0,\beta \to \alpha} = \mu_i^0(\alpha) - \mu_i^0(\beta). \tag{2.2.5}$$

The standard Gibbs transfer energy can only be found thermodynamically, for example, by using distribution equilibria, both for nonelectrolytes and for electrolytes as a whole. The distribution coefficient for substance X between phases α and β is given by the equation

$$k_X^{\beta,\alpha} = \exp\left(\frac{\mu_X^0(\alpha) - \mu_X^0(\beta)}{RT}\right) = \exp\left(\frac{\Delta G_{tr,X}^{0,\beta \to \alpha}}{RT}\right). \tag{2.2.6}$$

For the distribution equilibrium of uni–univalent electrolyte BA we have

$$k_{BA}^{\beta,\alpha} = \frac{a_\pm(\alpha)}{a_\pm(\beta)} = \frac{a_{B^+}(\alpha)a_{A^-}(\alpha)}{a_{B^+}(\beta)a_{A^-}(\beta)}. \tag{2.2.7}$$

If (2.2.4) is written for cation B^+ and anion A^-, then it follows from equality of the right-hand sides according to (2.2.5), (2.2.6) and (2.2.7) that

$$\Delta G_{tr,BA}^{0,\beta \to \alpha} = \tfrac{1}{2}(\Delta G_{tr,B^+}^{0,\beta \to \alpha} + \Delta G_{tr,A^-}^{0,\beta \to \alpha}). \tag{2.2.8}$$

No purely thermodynamic approach can be used to divide the experimentally determined value of the standard Gibbs energy for transfer of electrolyte BA as a whole into the contributions of the cation and the anion. Several extra-thermodynamic assumptions have been introduced to make this division possible. The most satisfactory has been found to be the TATB assumption introduced by Parker [29], according to which the standard Gibbs energy for the transfer of the tetraphenylarsonium ion (TPAs$^+$) and of the tetraphenylborate ion (TPB$^-$) are equal for any pair of solvents. The standard Gibbs energy for transfer of the TPAs$^+$ and TPB$^-$ ions can be determined from the distribution coefficients of the TPAsTPB salt between any pair of immiscible solvents. The distribution coefficient for the TPAsA salt is found for an arbitrary ion A^-, then its standard Gibbs transfer energy

$$\Delta G_{tr,A}^{0,\beta \to \alpha} = 2\Delta G_{tr,TPAsA}^{0,\beta \to \alpha} - \Delta G_{tr,TPAs}^{0,\beta \to \alpha}. \tag{2.2.9}$$

Equation (2.2.5) enables expression of quantity $\Delta G_{tr,i}^{0,\beta\to\alpha}$, which appears in (2.2.4), in electrical units according to the relationship [6]

$$\Delta G_{tr,i}^{0,\beta\to\alpha} = -z_i F \Delta_\beta^\alpha \phi_i^0,\qquad(2.2.10)$$

where $\Delta_\beta^\alpha \phi_i^0$ is the standard potential difference between phases α and β for ion i, a quantity analogous to the standard electrode potential. Substitution of (2.2.5) and (2.2.10) into (2.2.4) leads to a form analogous to the Nernst equation for the electrode potential

$$\Delta_\beta^\alpha \phi = \Delta_\beta^\alpha \phi_i^0 + (RT/z_i F)\ln\,[a_i(\beta)/a_i(\alpha)].\qquad(2.2.11)$$

The values of the standard Gibbs energy between nitrobenzene and water and the values of the standard potential differences between these two solvents for various ions are given in table 2.1. Equations (2.2.8) and (2.2.10) can be generalized readily for an electrolyte of any type of charge.

If a single uni–univalent electrolyte is in equilibrium between two immiscible solvents, then (2.2.11) for cation B^+ and anion A^-, assuming $a_{B^+}(\alpha)=a_{A^-}(\alpha)$ and $a_{B^+}(\beta)=a_{A^-}(\beta)$, can be combined with (2.2.10) to yield a relationship for the distribution potential [15]:

$$\Delta_\beta^\alpha \phi_{dstrb} = \tfrac{1}{2}(\Delta_\beta^\alpha \phi_{B^+}^0 + \Delta_\beta^\alpha \phi_{A^-}^0)$$
$$= (1/2F)(\Delta G_{tr,A^-}^{0,\beta\to\alpha} - \Delta G_{tr,B^+}^{0,\beta\to\alpha}).\qquad(2.2.12)$$

Generally, for a salt of cation B^{z+} and anion A^{z-},

$$\Delta_\beta^\alpha \phi_{dstrb} = \tfrac{1}{2}(-z_-\Delta_\beta^\alpha \phi_{B^+}^0 + z_+\Delta_\beta^\alpha \phi_{A^-}^0)$$
$$= (1/2F)[(z_-/z_+)\Delta G_{tr,B^+}^{0,\beta\to\alpha} - (z_+/z_-)\Delta G_{tr,A^-}^{0,\beta\to\alpha}].\qquad(2.2.13)$$

The physical sense of the distribution potential can be demonstrated on the example of the distribution equilibrium of the salt of a hydrophilic cation and a hydrophobic anion between water (wt) and an organic solvent that is immiscible with water (org). After attaining distribution equilibrium the concentrations of the anion and the cation in each of the two phases are the same because of the electroneutrality condition. However, at the phase boundary an electrical double layer is formed as a result of the greater tendency of the anions to pass from the aqueous phase into the organic phase, and of the cations to move in the opposite direction. This can be characterized quantitatively by quantities $-\Delta G_{tr,B^+}^{0,org\to wt}$ and $-\Delta G_{tr,A^-}^{0,org\to wt}$, for which

$$-\Delta G_{tr,A^-}^{0,org\to wt} < -\Delta G_{tr,B^+}^{0,org\to wt}.\qquad(2.2.14)$$

It follows from this relationship and from (2.2.12) that $\Delta_{org}^{wt}\phi_{dstrb} > 0$, in agreement with the concept that the part of the double layer facing the aqueous phase is positively charged, while the part in the organic solvent is negatively charged.

The general case of a distribution potential for an arbitrary number of salts, for the formation of ion pairs and for complex formation was solved by Hung [12].

2.3   The Donnan potential [5]

Consider a system containing two electrolytes separated by a membrane preventing transfer of one kind of ion from one phase into the other, so that this ion is present only in one phase even when the system is in equilibrium. A suitable model is a microporous membrane through which an ion of high molecular weight cannot pass.

Table 2.1. *Values of standard Gibbs energies of ion transfer from water to nitrobenzene in electron volts. From P. Vanýsek, Thesis, J. Heyrovský Institute of Physical Chemistry and Electrochemistry, Czechoslovak Academy of Sciences, Prague (1982).*

Ion	$G_{tr,i}^{0,nb \to wt}$ /kJ mol	$\Delta_{nb}^{wt} \phi_i^0$ /mV
Li^+	-38.2	395
Na^+	-34.2	354
Ca^{2+}	-67.3	349
H^+	-32.5	337
NH_4^+	-26.8	277
K^+	-23.4	242
Rb^+	-19.4	201
Cs^+	-15.4	159
Choline$^+$	-11.3	117
Acetylcholine$^+$	-4.8	49
TMeA$^+$ (tetramethylammonium)	-3.4	35
TEA$^+$ (tetraethylammonium)	5.7	-59
TBA$^+$ (tetrabutylammonium)	24.0	-248
TPAs$^+$ (tetraphenylarsonium)	35.9	-372
Crystal Violet$^+$	39.5	-410
Dicarbolylcobaltate$^-$	50.2	520
Dipicrylaminate	39.4	407
I_5^-	23.4	401
TPB$^-$ (tetraphenylborate)	35.9	372
I_3^-	23.4	242
Octanate	8.5	89
Picrate	4.6	47
Dodecylsulfate	-4.1	-43
SCN$^-$	-5.8	-61
ClO_4^-	-8.0	-83
I^-	-18.8	-195
Br^-	-28.4	-295
Cl^-	-31.4	-324
F^-	-44.0	-454

In the simplest case, such a system consists of two phases with the same volume. One of them (phase 1) contains KCl in concentration c_1 and the potassium salt of a macromolecular anion, KX, in concentration c_X. The second phase consists only of a KCl solution with concentration c_2. The system can be depicted by the scheme

$$\text{Cl}^-(c_1), \text{K}^+(c_1 + c_X), \text{X}^-(c_X) \,\|\, \text{Cl}^-(c_2), \text{K}^+(c_2) \qquad (2.3.1)$$

$$\underset{\text{phase 1}}{} \qquad\qquad \underset{\text{phase 2}}{}$$

At equilibrium, a certain amount of KCl must pass from one phase into the other in order that the equilibrium characterized for the K^+ and Cl^- ions (termed diffusible ions) by (2.2.2) may be established, i.e.

$$\begin{aligned} \tilde{\mu}_{\text{K}^+}(1) &= \tilde{\mu}_{\text{K}^+}(2) \\ \tilde{\mu}_{\text{Cl}^-}(1) &= \tilde{\mu}_{\text{Cl}^-}(2). \end{aligned} \qquad (2.3.2)$$

For dilute solutions, $(a_{\text{K}^+} \approx c_{\text{K}^+}, a_{\text{Cl}^-} \approx c_{\text{Cl}^-})$

$$c_{\text{K}^+}(1) c_{\text{Cl}^-}(1) = c_{\text{K}^+}(2) c_{\text{Cl}^-}(2), \qquad (2.3.3)$$

where

$$\begin{aligned} c_{\text{K}^+}(1) &= c_1 + c_X - y, \\ c_{\text{Cl}^-}(1) &= c_1 - y, \\ c_{\text{K}^+}(2) &= c_2 + y, \\ c_{\text{Cl}^-}(2) &= c_2 + y. \end{aligned}$$

After substitution into (2.3.3) a relationship is obtained for concentration y, by which the original concentrations of KCl, c_1 and c_2, must change.

It should be noted that the condition of a dilute solution was introduced into the considerations for two reasons: primarily, in order that it would be possible to replace the activities by concentrations and thus determine the equilibrium concentrations on the basis of (2.3.3); and, secondarily, in order for it to be possible to neglect the effect of pressure on the chemical potentials of the components whose electrochemical potentials appear in (2.3.2). Because of the differing ionic concentrations in solutions 1 and 2, the osmotic pressures in these solutions are not identical and this difference must be compensated by external pressure. A derivation considering the effect of pressure can be found, for example in [9] or p. 191 of [18].

2.4 The structure of the phase boundary between two immiscible electrolytes

As mentioned in section 2.2, an electric double layer is formed at the boundary between two immiscible electrolyte solutions as a result of the different tendencies of electrically-charged components in each of the phases to pass into the other phase.

At the phase boundary between water and an organic solvent immiscible with water, an electric double layer may also be formed if one of the solutions in contact contains an amphiphilic ion, i.e. an ion that contains both hydrophilic (usually electrically charged) and hydrophobic (nonpolar) groups (for example the dodecyltrimethylammonium ion). Ions with these properties are simply adsorbed on the phase boundary with the nonpolar group oriented into the organic solvent and the ionized group in the aqueous phase.

Similar types of electric double layer may also be formed at the phase boundary between a solid electrolyte and an aqueous electrolyte solution [7]. They are formed because one electrically-charged component of the solid electrolyte is more readily dissolved, for example the fluoride ion in solid LaF_3, leading to excess charge in the solid phase, which, as a result of movement of the holes formed, diffuses into the solid electrolyte. Another possible way a double layer may be formed is by adsorption of electrically-charged components from solution on the phase boundary, or by reactions of such components with some component of the solid electrolyte. For LaF_3 this could be the reaction of hydroxyl ions with the trivalent lanthanum ion. Characteristically, for the phase boundary between two immiscible electrolyte solutions, where neither solution contains an amphiphilic ion, the electric double layer consists of two diffuse electric double layers, with no compact double layer at the actual phase boundary, in contrast to the metal electrode/ electrolyte solution boundary [4, 8, 35] (see fig. 2.1). Then, for the potential

Fig. 2.1. Distribution of the electric potential ϕ in the vicinity of ITIES. The position of ITIES is indicated by x_0, cf. (2.4.1).

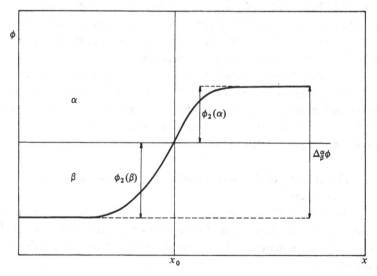

difference between the two phases,

$$\Delta_\beta^\alpha \phi = \phi_2(\alpha) + \phi_2(\beta), \tag{2.4.1}$$

where $\phi_2(\alpha)$ is the potential difference between the inside of phase α and the phase boundary, and $\phi_2(\beta)$ is the potential difference between the phase boundary and the inside of phase β. As a whole, the electric double layer is electroneutral; so for the electric charge in the part of the double layer in phase α, $q(\alpha)$, and for the part in phase β, $q(\beta)$,

$$q(\alpha) + q(\beta) = 0. \tag{2.4.2}$$

According to the Gouy-Chapman theory of the diffuse electric double layer (chapter 3 of [18]), for a uni–univalent electrolyte,

$$q(\alpha) = [8\epsilon(\alpha)RTc(\alpha)]^{1/2} \sinh [F\phi_2(\alpha)/2RT],$$
$$q(\beta) = -[8\epsilon(\beta)RTc(\beta)]^{1/2} \sinh [F\phi_2(\beta)/2RT], \tag{2.4.3}$$

where $\epsilon(\alpha)$ and $\epsilon(\beta)$ are the permittivities of phases α and β respectively, and $c(\alpha)$ and $c(\beta)$ are the electrolyte concentrations in phases α and β respectively. The chemical composition of the electrolyte is not important. The same electrolyte may be present in both phases if distribution equilibrium is established between the two phases. On the other hand, if the standard Gibbs transfer energies from phase β into phase α of both components present in phase α have large negative values, and if both components of another electrolyte in the phase β have large positive values, then in practice one electrolyte is present only in phase α and the other only in phase β. Under these conditions, the phase boundary behaves in a similar way to an ideally polarized electrode [19]. In spite of the differences between these two situations, it is possible to find a simple relationship between the individual values of $\phi_2(\alpha)$ and $\phi_2(\beta)$, the overall potential difference $\Delta_\beta^\alpha \phi$ and the quantities $\epsilon(\alpha)$, $\epsilon(\beta)$, $c(\alpha)$ and $c(\beta)$. It follows from (2.4.2) and (2.4.3) that

$$\frac{\sinh (F\phi_2(\alpha)/2RT)}{\sinh (F\phi_2(\beta)/2RT)} = [\epsilon(\alpha)c(\alpha)/\epsilon(\beta)c(\beta)]^{1/2}. \tag{2.4.4}$$

$\phi_2(\alpha)$ and $\phi_2(\beta)$ can be calculated from this relationship and from (2.4.2). For $|\Delta_\beta^\alpha \phi| \gg RT/F$,

$$\phi_2(\alpha) = \tfrac{1}{2}\Delta_\beta^\alpha \phi \pm (RT/2F) \ln [\epsilon(\alpha)c(\alpha)/\epsilon(\beta)c(\beta)]$$
$$\phi_2(\beta) = \tfrac{1}{2}\Delta_\beta^\alpha \phi \mp (RT/2F) \ln [\epsilon(\alpha)c(\alpha)/\epsilon(\beta)c(\beta)], \tag{2.4.5}$$

where the plus sign holds for $\Delta_\beta^\alpha \phi > 0$ and the minus sign for $\Delta_\beta^\alpha \phi < 0$. These relationships completely describe the potential distribution at the boundary between two immiscible electrolyte solutions and so also at the liquid membrane/electrolyte solution boundary, assuming that the Gouy-Chapman theory and (2.4.1) are valid.

2.5   The kinetics of charge transfer across the phase boundary between two electrolytes

Consider a system in which a potential difference ΔV, in general different from the equilibrium potential between the two phases $\Delta_\beta^\alpha \phi$, is applied from an external source to the phase boundary between two immiscible electrolyte solutions. Then an electric current is passed, which in the simplest case corresponds to the transfer of a single kind of ion across the phase boundary. Assume that the Butler–Volmer equation for the rate of an electrode reaction (see p. 255 of [18]) can also be used for charge transfer across the phase boundary between two electrolytes (cf. [16, 19]). It is mostly assumed (in the framework of the Frumkin correction) that only the potential difference in the compact part of the double layer affects the actual charge transfer, so that it follows for the current density in our system that

$$j = z_i F k_i^0 \{c_i^*(\alpha) \exp [\alpha' z_i F(\Delta V - \phi_2(\alpha) - \phi_2(\beta) - \Delta_\beta^\alpha \phi_i^0)/RT]$$
$$-c_i^*(\beta) \exp [-(1-\alpha')z_i F(\Delta V - \phi_2(\alpha) - \phi_2(\beta) - \Delta_\beta^\alpha \phi_i^0)/RT]\},$$

$$(2.5.1)$$

where z_i is the charge number of the ion passing across the phase boundary, $\Delta_\beta^\alpha \phi_i^0$ is the standard potential difference between phases α and β for this ion, k_i^0 is the standard rate constant for transfer of ion i and α' is the charge-transfer coefficient. Concentrations $c_i^*(\alpha)$ and $c_i^*(\beta)$ correspond to the immediate vicinity of the phase boundary and are functions of the potential differences in the diffuse double layers according to the Boltzmann relationship

$$c_i^*(\alpha) = c_i(\alpha) \exp [-z_i F\phi_2(\alpha)/RT]$$
$$c_i^*(\beta) = c_i(\beta) \exp [z_i F\phi_2(\beta)/RT].$$

$$(2.5.2)$$

Theoretical analysis of (2.5.1) based on the Marcus–Levich charge-transfer theory can be found in [34].

If it is assumed that a relationship analogous to (2.4.1) holds for ΔV even during passage of current across the boundary between two immiscible solutions,

$$\Delta V = \phi_2(\alpha) + \phi_2(\beta) \qquad (2.5.3)$$

then (2.5.2) and (2.5.3) can be used to reduce (2.5.1) to the form

$$j = z_i F k_i^0 \{c_i(\alpha) \exp (-\alpha' z_i F \Delta_\beta^\alpha \phi_i^0/RT) \exp [z_i F\phi_2(\alpha)/RT]$$
$$-c_i(\beta) \exp [(1-\alpha')z_i F \Delta_\beta^\alpha \phi_i^0/RT] \exp [-z_i F\phi_2(\beta)/RT]\}. \quad (2.5.4)$$

Formal validity of (2.5.4) follows from the fact that at equilibrium ($j = 0$), (2.5.4) is converted into (2.2.11), assuming that $a_i \approx c_i(\Delta V = \Delta_\beta^\alpha \phi)$.

Consider a system in which both phases further contain uni–univalent electrolytes in concentrations of $c(\alpha)$ and $c(\beta)$, where the ion taking part in charge transfer is in a much lower concentration than these indifferent electrolytes. Then, for $|\Delta V| \gg RT/F$, (2.5.4) becomes

$$j = z_i F k_i^0 \left[\epsilon(\alpha)c(\alpha)/\epsilon(\beta)c(\beta)\right]^{\pm z_i/2} \exp\left[(\tfrac{1}{2} - \alpha')z_i F/(RT)\Delta_\beta^\alpha \phi_i^0\right]$$
$$\times \{c_i(\alpha) \exp\left[z_i F/(2RT)(\Delta V - \Delta_\beta^\alpha \phi_i^0)\right]$$
$$- c_i(\beta) \exp\left[-z_i F/(2RT)(\Delta V - \Delta_\beta^\alpha \phi_i^0)\right]\}, \quad (2.5.5)$$

where the plus sign in the exponent of the first term in square brackets is valid for $\Delta V > 0$, while the minus sign is valid for $\Delta V < 0$. It is apparent that, assuming (2.5.3) is valid, (2.5.1) is simplified to the simple Butler–Volmer equation with $\alpha = \tfrac{1}{2}$ [17, 23, 35]. The rate of ion transfer depends markedly on the ratio of the dielectric constants and on the concentrations of indifferent electrolytes in the two phases, as well as on the distribution coefficient for ion i, given (2.2.6) and (2.2.10),

$$k_i^{\beta,\alpha} = \exp\{[z_i F/(RT)]\Delta_\beta^\alpha \phi_i^0\}. \quad (2.5.6)$$

Rate constant k_i^0 (dimension cm s^{-1}) is given by the product

$$k_i^0 P \exp(-E_r/RT), \quad (2.5.7)$$

where P is the frequency factor and E_r the resolution energy needed for the change of the solvation sheath in one solvent to the solvation sheath in the other solvent. The frequency factor P can be interpreted as the ratio of the distance between a site adjacent to the phase boundary in phase α, at which species i is located just before passing to a similar site in phase β, and time τ during which this transfer can occur. The distance between these two sites is approximately equal to the sum of the diameters of molecules of solvents α and β. According to Einstein and Smoluchowski, for the shift of a species in an arbitrary direction Δx,

$$\Delta x = \tfrac{1}{2}(2D\tau)^{1/2}, \quad (2.5.8)$$

where D is the mean diffusion coefficient of the species, to a first approximation equal to the geometric average of the diffusion coefficients in the two phases $(D_\alpha D_\beta)^{1/2}$.

For the ratio of the mean shift and the corresponding time τ,

$$\Delta x/\tau = (D_\alpha D_\beta)^{1/2}/2\Delta x = (D_\alpha D_\beta)^{1/2}/2(r_\alpha + r_\beta). \quad (2.5.9)$$

As the species can move by distance Δx in time τ in any direction, it holds for a shift in the direction perpendicular to the boundary that $\Delta x' = 1/6\Delta x$ and thus for the rate constant that

$$k_i^0 = \frac{\Delta x'}{\tau} \exp(-E_r/RT) = \frac{(D_\alpha D_\beta)^{1/2}}{12(r_\alpha + r_\beta)} \exp(-E_r/RT) \quad (2.5.10)$$

No theory has been worked out for E_r so far.

2.6 Diffusion potential

The membranes that will be considered in this book are frequently not in a state of equilibrium. Consequently, component transport occurs and, with charged species, also charge transfer. The rate of transport of an arbitrary component of the given system is characterized by the mass flux J_i (the amount of substance passing through a unit area per unit time). This flux is proportional to the mobility of the species, U_i, to its concentration at a given site, c_i, and to the driving force for the transport process. The species mobility has dimension of kg^{-1} s mol in basic SI units; for charge transport the electrolytic mobility for an ion with charge number z_i

$$u_i = |z_i| FU_i$$

is used, with the basic unit $m^2 s^{-1} V^{-1}$. For the electrolytic mobility, $u_i = \lambda_i/F$, where λ_i is the ionic conductivity. Mobility U_i is then

$$U_i = \lambda_i/|z_i| F^2.$$

If the effect of external mechanical forces on the system is negligible (i.e. if convection does not occur), the only driving force for a transport process is the electrochemical potential gradient. For a change in a system occurring only along the x-coordinate,

$$J_i = -U_i c_i \, d\bar{\mu}_i/dx. \tag{2.6.1}$$

In a dilute solution the activities can be replaced by concentrations and (2.6.1) yields the Nernst–Planck equation

$$J_i = -U_i RT \, dc_i/dx - z_i Fu_i c_i \, d\phi/dx. \tag{2.6.2}$$

It follows from comparison of this equation with Fick's first law that

$$U_i RT = D_i, \tag{2.6.3}$$

where D_i is the diffusion coefficient of component i.

A simple system in which transport processes occur that are also characteristic of membrane processes is the liquid junction formed between two electrolyte solutions in the same solvent. The region in which one electrolyte passes into the other is frequently a porous diaphragm of various construction (fig. 2.2). A second type of liquid junction is the 'free diffusion' region (fig. 2.3).

The only process occurring in a liquid junction is the diffusion of various components of the two solutions in contact with it. The various mobilities of the ions present in the liquid junction lead to the formation of an electric potential gradient, termed the diffusion potential gradient. A potential difference, termed the liquid-junction potential, $\Delta\phi_L$, is formed between two solutions whose composition is assumed to be constant outside the liquid junction.

Mass flux of the electrically charged component J_i is connected with charge transport and so with the partial current density j_i defined as

$$j_i = z_i F J_i. \tag{2.6.4}$$

If no electric current passes through the liquid junction, then the overall current density

$$j = \sum j_i = 0. \tag{2.6.5}$$

Substitution of (2.6.2) and (2.2.1) into (2.6.4) yields, on rearrangement,

$$\frac{d\phi}{dx} = -\frac{RT}{F} \frac{\sum\limits_i z_i U_i \, dc_i/dx}{\sum\limits_j z_j^2 U_j c_j} = -\frac{RT}{F} \frac{\sum\limits_i z_i U_i c_i \, d\ln c_i/dx}{\sum\limits_j z_j^2 U_j c_j}$$

$$= -\frac{RT}{F} \sum \frac{t_i}{z_i} \frac{d\ln c_i}{dx}, \tag{2.6.6}$$

Fig. 2.2. Various types of liquid junction with restrained diffusion: (a) ceramic plug; (b) groundglass jacket; (c) asbestos wick; (d) U-shaped connection with a ceramic plug. (From G. Mattock and D. M. Band, Interpretation of pH and cation measurement, chapter 2 of *Glass Electrodes for Hydrogen and Other Cations* (ed. G. Eisenman), M. Dekker, New York (1967). By permission of M. Dekker Inc.)

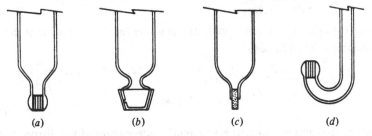

(a)　　　　(b)　　　　(c)　　　　(d)

Fig. 2.3. Liquid junction with free diffusion: (a) the test solution is drawn (in the direction of the arrow) into the stopcock; (b) the saturated KCl solution from the liquid bridge is drawn into the stopcock; (c) by turning the stopcock the test solution/liquid-bridge liquid junction is formed. (After Mattock and Band.)

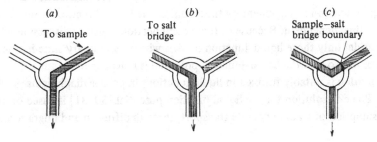

where t_i is the transport number of the ith component,

$$t_i = z_i^2 U_i c_i / \sum_j z_j^2 U_j c_j. \tag{2.6.7}$$

The liquid-junction potential is [26, 27]

$$\Delta\phi_L = \phi(2) - \phi(1) = -RT/F \int_1^2 \frac{t_i}{z_i} \, d \ln c_i, \tag{2.6.8}$$

where $\phi(2)$ and $\phi(1)$ are the electric potentials of solutions 2 and 1 respectively, with which the liquid junction is in contact, and the limits of the integral in the last term are related to the compositions of these solutions. It should be noted that concentrations must be replaced by activities in the rigorous formulation.

The value of $\Delta\phi_L$ depends in general on the nature of functions x that describe the concentration distribution c_i and thus also on quantity t_i. The simplest situation occurs when there are solutions of a single electrolyte at different concentrations c_1 and c_2 on the two sides of the liquid junction. For the transport number,

$$t_+ = \frac{z_+ U_+}{z_+ U_+ - z_- U_-}$$
$$t_- = \frac{-z_- U_-}{z_+ U_+ - z_- U_-}, \tag{2.6.9}$$

where the subscripts $+$ and $-$ indicate the values for the cation and anion, respectively. Then for $\Delta\phi_L$,

$$\Delta\phi_L = -\frac{RT}{F}\left(\frac{t_+}{z_+} + \frac{t_-}{z_+}\right) \ln \frac{c_2}{c_1} = -\frac{RT}{F}\left(\frac{z_+ - (z_+ - z_-)t_+}{z_+ z_-}\right) \ln \frac{c_2}{c_1}. \tag{2.6.10}$$

The liquid–junction potential is apparently independent of the distribution of ion concentrations in this system.

In the general case of several electrolytes present in the solutions in contact with the liquid junction, no simple result analogous to (2.6.10) can be obtained. A basic problem stems from the fact that the electrolyte distribution in the liquid junction is dependent on time, so that the liquid–junction potential is also time-dependent. Because of these complications, further discussion will consider only those liquid–junction models where a stationary state has been attained, so that the liquid–junction potential is independent of time. This condition is notably fulfilled in liquid junctions in porous diaphragms.

Planck's solution for the liquid junction potential [30, 31] is based on the assumption of stationary state transport, through diffusion and migration, and of identical absolute values of charge numbers for the cation and anion. As will be seen in the next chapter, considerable complications are encountered, even under some other conditions, if this latter condition is not fulfilled. Planck's solution leads to a rather complicated result, which cannot be given in closed form. His procedure is reproduced in MacInnes' textbook of electrochemistry [21] and in the first edition of this book. Recently, a simpler form of the Planck procedure was developed by Morf [24] (see also [16]).

In contrast to the Planck solution, the Henderson approximation enjoys considerable use [10, 11]. Henderson's liquid–junction model is based on the assumption that the concentrations of the ions in the liquid junction change linearly with x between values corresponding to the edges of the liquid junction. This assumption is equivalent to the concept of a mixture of electrolytes changing uniformly between the two edges of the liquid junction. Then

$$c_i(x) = c_i(0) + [c_i(d) - c_i(0)]x/d, \qquad (2.6.11)$$

where the coordinate values $x = 0$ and $x = d$ correspond to the left and right edges of the liquid junction in the usual graphical form. Substituting (2.6.11) into (2.6.6) yields

$$\Delta\phi_L = \phi(2) - \phi(1)$$

$$= -\frac{RT}{F} \int_0^d \frac{\sum_i z_i U_i \{[c_i(d) - c_i(0)]x/d + c_i(0)\}}{\sum_j z_j^2 U_j \{[c_i(d) - c_i(0)]x/d + c_j(0)\}}$$

$$\times \, d \ln \{[c_i(d) - c_i(0)]x/d + c_i(0)\}$$

$$= -\frac{RT}{F} \frac{\sum_i z_i U_i [c_i(d) - c_i(0)]}{\sum_i z_i^2 U_i [c_i(d) - c_i(0)]} \ln \frac{\sum_i z_i^2 U_i c_i(d)}{\sum_i z_i^2 U_i c_i(0)}. \qquad (2.6.12)$$

While the Planck liquid–junction model corresponds to a junction with 'restrained flow', for example in a porous diaphragm, fig. 2.2, the Henderson model approaches a liquid junction with 'free diffusion' (fig. 2.3). Ives and Janz [13] give inaccuracies in measuring liquid–junction potentials between 1 and 2 mV.

If a single uni–univalent electrolyte with a common cation or anion is present in the same concentration on both sides of the liquid junction, for example in the arrangement

0.1 M KCl ‖ 0.1 M HCl

solution 1 ‖ solution 2

then (2.6.12) is reduced, as given by Lewis and Sargent [20], to the form

$$\Delta\phi_L = \mp \frac{RT}{F} \ln \frac{U_+(2) + U_-(2)}{U_+(1) + U_-(1)} = \mp \frac{RT}{F} \ln \frac{\Lambda(2)}{\Lambda(1)}, \qquad (2.6.13)$$

where symbols 1 and 2 designate the electrolytes in solutions 1 and 2 and $\Lambda(1)$ and $\Lambda(2)$ are the equivalent conductivities at the given electrolyte concentration. The minus sign is valid when both electrolytes have a common anion, while the plus sign is valid for a common cation. As follows from table 2.2 [22], the experimental data agree very well with the theory [20].

A very general solution of (2.6.8), which is, however, not very useful in practice, was given by Pleijel [32] and Schlögl [36]. A comparison of the liquid–junction potential values calculated by various methods is given in the book by Morf [25]. Comparison shows that the Henderson equation (2.6.12) is satisfactory for practical measurements in spite of its theoretical weaknesses. Only when there is a quite concentrated solution of a strong acid on one side of the liquid junction do large errors arise as a result of the presence of the very mobile hydrogen ion in a large concentration. A similar effect may be observed to a lesser degree in the presence of a quite concentrated solution of a strong base on one side of the liquid junction.

Ideal potentiometric measurements, especially in analytical chemistry, would require that the potential of the reference electrode be fixed and known, and that the composition of the studied solution affect only the potential of the indicator electrode. This would occur only if the liquid–junction potential could be completely neglected. In practice this situation can be attained only if the whole system contains an indifferent electrolyte in a much larger concentration than that of the other electrolytes, so that the concentration of a particular component in the analysed solution, which is not present in the reference electrode solution, has only a negligible effect on the liquid–junction potential. Such a situation rarely occurs, so that it is necessary to know or at least fix the liquid junction potential.

Table 2.2. *Liquid-junction potential values; all values at 25 °C according to (2.6.13) compared with experiment* [22].

Liquid junction			
Electrolyte 2	Electrolyte 1	$\Delta\phi_L$ (exp.)/mV	$\Delta\phi_L$ (calc.)/mV
0.1 M HCl	0.1 M KCl	−26.78	−28.52
0.1 M HCl	0.1 M KCl	−33.09	−33.38
0.1 M KCl	0.1 M NaCl	−6.42	−4.86
0.1 M NaCl	0.1 M NH$_4$Cl	+4.21	+4.81
0.01 M HCl	0.01 M KCl	−25.73	−27.48
0.01 M KCl	0.01 M NaCl	−5.65	−4.54

An acceptable method quite frequently used in practice depends on the cell whose EMF is being measured having a liquid junction with a constant potential value. Such a situation is attained in the determination of the activity of fluoride ions, by adding a constant amount of quite concentrated buffer, for example TISAB, to the studied solution; this buffer also fulfills other functions in the analysis (see p. 146). Then the liquid junction potential is a function of the composition of the reference electrode electrolyte and of the buffer composition alone, and not of the concentrations of the other components of the studied solution.

Another less precise but frequently used method employs a liquid bridge between the analysed solution and the reference electrode solution. This bridge is usually filled with a saturated or 3.5 M KCl solution. If the reference electrode is a saturated calomel electrode, no further liquid bridge is necessary. Use of this bridge is based on the fact that the mobilities of potassium and chloride ions are about the same so that, as follows from the Henderson equation, the liquid–junction potential with a dilute solution on the other side has a very low value. Only when the saturated KCl solution is in contact with a very concentrated electrolyte solution with very different cation and anion mobilities does the liquid junction potential attain larger values [2]; for the liquid junction 3.5 M KCl ‖ 1 M NaOH, $\Delta\phi_L = 10.5$ mV.

In biological measurements, liquid bridges filled with agar gel saturated with KCl are advantageous. A similar bridge filled with saline (0.9% NaCl) is also frequently used. As the mobilities of sodium and chloride ions are different ($t_{Na^+} = 0.40$), a quite large liquid–junction potential is formed at the liquid bridge/test solution boundary, which depends on the composition of the studied solution. If it is necessary, for example as in measurements on a cellular level, that the liquid bridge does not contain potassium ions, it is suitable to use sodium formate ($t_{Na^+} = 0.48$) as an electrolyte.

In the presence of colloidal solutions in contact with a liquid junction, anomalous liquid–junction potentials are often measured. This suspension or Palmann effect [14] has not yet been satisfactorily explained. It is probably a Donnan-type potential with the electrically-charged colloidal species acting as indiffusible ions (cf. section 5.1.3).

References for chapter 2

1 K. F. Bonhoeffer, M. Kahlweit and H. Strehlow, *Z. Elektrochemie* 57, 614 (1953).
2 R. P. Buck, Potentiometry: pH measurements and ion selective electrodes, chapter 2 of *Physical Methods of Chemistry* (ed. A. Weissberger and B. W. Rossiter), Part IIA, Interscience, New York (1971).

3 R. P. Buck, Theory and principles of membrane electrodes, chapter 1 of *Ion-Selective Electrodes in Analytical Chemistry* (ed. H. Freiser), Plenum Press, New York (1978).
4 B. D'Epenoux, P. Seta, G. Amblard and C. Gavach, *J. Electroanal. Chem.* **94**, 77 (1979).
5 F. G. Donnan, *Z. Elektrochemie.* **17**, 572 (1911).
6 C. Gavach and F. Henry, *J. Electroanal. Chem.* **54**, 361 (1974).
7 T. B. Grimley and N. F. Mott, *Faraday Soc. Discussions* **1**, 9 (1947).
8 M. Gros, S. Gromb and C. Gavach, *J. Electroanal. Chem.* **89**, 29 (1978).
9 E. A. Guggenheim, *Thermodynamics*, North Holland, Amsterdam (1949).
10 P. Henderson, *Z. Physik. Chem.* **59**, 118 (1907).
11 P. Henderson, *Z. Physik. Chem.* **63**, 325 (1908).
12 Le Q. Hung, *J. Electroanal. Chem.* **115**, 159 (1980).
13 D. J. G. Ives and G. J. Janz, *Reference Electrodes*, Academic Press, New York (1961).
14 H. Jenny, T. R. Clark, N. T. Coleman and D. E. Williams, *Science* **112**, 164 (1950).
15 F. M. Karpfen and J. E. B. Randles, *Trans. Faraday Soc.* **49**, 823 (1953).
16 J. Koryta, *Anal. Chim. Acta* **111**, 1 (1979).
17 J. Koryta, *Anal. Chim. Acta* **139**, 1 (1982).
18 J. Koryta, J. Dvořák and V. Boháčková, *Electrochemistry*, Chapman & Hall, London (1973).
19 J. Koryta, P. Vanýsek and M. Březina, *J. Electroanal. Chem.* **75**, 211 (1977).
20 G. N. Lewis and L. W. Sargent, *J. Am. Chem. Soc.* **31**, 363 (1909).
21 D. A. MacInnes, *The Principles of Electrochemistry*, Dover, New York (1939).
22 D. A. MacInnes and Y. L. Yeh, *J. Am. Chem. Soc.* **43**, 2563 (1921).
23 O. R. Melroy and R. P. Buck, *J. Electroanal. Chem.* **136**, 19 (1982).
24 W. E. Morf, *Anal. Chem.* **49**, 810 (1977).
25 W. E. Morf, *The Principles of Ion-Selective Electrodes and of Membrane Transport*, Akadémiai Kiadó, Budapest (1981).
26 W. Nernst, *Z. Physik. Chem.* **2**, 613 (1889).
27 W. Nernst, *Z. Physik. Chem.* **4**, 129 (1889).
28 W. Nernst and E. H. Riesenfeld, *Ann. Phys.* **8**, 600 (1902).
29 A. J. Parker, *Electrochim. Acta*, **21**, 671 (1976).
30 M. Planck, *Ann. Phys. Chem.* N. F. **39**, 196 (1890).
31 M. Planck, *Ann. Phys. Chem.* N. F. **40**, 561 (1890).
32 H. Pleijel, *Z. Physik. Chem.* **72**, 1 (1910).
33 E. H. Riesenfeld, *Ann. Phys.* **8**, 609 (1902).
34 Z. Samec, *J. Electroanal. Chem.* **99**, 197 (1979).
35 Z. Samec, V. Mareček and D. Homolka, *J. Electroanal. Chem.* **126**, 121 (1981).
36 R. Schlögl, *Z. Physik. Chem.* **1**, 305 (1954).

3

THE THEORY OF MEMBRANE POTENTIALS
IN ION-SELECTIVE ELECTRODES

3.1 Basic concepts

In potentiometry with ion-selective electrodes (ISE), as a rule, it is the electromotive force of a cell such as the following that is measured.

| reference | solution 1 | membrane | solution 2 | reference |
| electrode 1 | | | | electrode 2 |

$$(3.1.1)$$

The actual ISE system consists of internal reference electrode 2, internal solution of the ion-selective electrode 2 and the membrane. This system is immersed in analysed solution (analyte) 1, in which reference electrode 1 is also immersed, usually connected through a liquid bridge. The electromotive force of cell (3.1.1) is

$$E = E_2 + \Delta\phi_M - E_1. \qquad (3.1.2)$$

Reference electrodes E_1 and E_2 are usually electrodes of the second kind; their potentials are tabulated or can readily be calculated. In the analytical use of ISEs, the membrane potential $\Delta\phi_M$ is of basic importance. The sum of the potential of the internal reference electrode and the membrane potential is termed the ISE potential, E_{ISE},

$$E_{ISE} = E_2 + \Delta\phi_M. \qquad (3.1.3)$$

If the ISE membrane is in contact on both sides with solutions of the same ion whose activity is to be determined in the analyte (determinand J), and neither the analysed nor the internal ISE solution contains another ion that would affect the membrane potential (interferent K), then the membrane potential is

$$\Delta\phi_M = (RT/zF) \ln [a_J(1)/a_J(2)], \qquad (3.1.4)$$

where z is the charge number of ion J and $a_J(1)$ and $a_J(2)$ are its activities in solutions 1 and 2. Equation (3.1.4) is identical with the Nernst equation for the potential of a concentration cell without transfer. The ISE potential is then

$$E_{\text{ISE}} = \text{const} + (RT/zF)\ln a_J(1),\tag{3.1.5}$$

where the constant is determined by quantities that are independent of the activity of the determinand in the analyte.

If the analyte contains interferent K with activity $a_K(1)$ and the same charge number as ion J, the ISE membrane potential is

$$\Delta\phi_M = \frac{RT}{zF}\ln\frac{a_J(1) + k_{J,K}^{\text{pot}}a_K(1)}{a_J(2)}\tag{3.1.6}$$

and the ISE potential is given by the Nikolsky equation [58]

$$\begin{aligned}E_{\text{ISE}} &= \text{const} + (RT/zF)\ln\left[a_J(1) + k_{J,K}^{\text{pot}}a_K(1)\right]\\ &= \text{const} + (2.3\,RT/zF)\log\left[a_J(1) + k_{J,K}^{\text{pot}}a_K(1)\right],\end{aligned}\tag{3.1.7}$$

where $k_{J,K}^{\text{pot}}$ is the selectivity coefficient of the ISE for ion J with respect to ion K.

It will be shown below that the selectivity coefficient is frequently a function of the composition of the analyte; nonetheless, it provides an indication of the effect of interfering ions on the determination of the test ion. It can be seen that the ISE potential yields the activity of the test component, not its concentration. The activity can, however, be used to calculate the concentration if the activity coefficient is known.

An ideal ISE would exhibit a specific response to a certain ion J and the effect of interferents would be excluded. Except for the silver sulphide electrode, which is specific for sulphide or silver ions, no ion-selective electrode has this property. The others exhibit selectivity only for a particular ion with respect to the others. The selective behaviour of an ISE follows from (3.1.7). If the activity of the interferent is sufficiently low, i.e. if

$$a_J(1) \gg k_{J,K}^{\text{pot}}a_K(1),\tag{3.1.8}$$

then (3.1.7) is converted into (3.1.5). Then the E_{ISE} versus $\log a_J(1)$ dependence has the slope

$$\frac{dE_{\text{ISE}}}{d\log a_K} = \frac{2.3\,RT}{zF}\tag{3.1.9}$$

called the Nernstian slope. When the interferent concentration is sufficiently high, so that

$$a_J(1) \ll k_{J,K}^{\text{pot}}a_K(1)\tag{3.1.10}$$

(3.1.7) is converted to an equation of the form (3.1.5), but this time for ion K, so that the ISE potential depends on the activity of ion K with Nernstian slope.

The Nikolsky equation (3.1.7) frequently corresponds to the E_{ISE} versus $\log a_J$ dependence only in those regions where (3.1.8) or (3.1.10) is valid and is thus often only an operational formula with the same asymptotes as the actual E_{ISE} versus $\log a_J$ dependence. Nonetheless, the selectivity coefficient obtained from (3.1.7) is important in estimating the ISE selectivity, although this equation cannot be used where the E_{ISE} versus $\log a_J$ dependence is clearly curved. In the intercept of the asymptotes to (3.1.7), $\log a_J$ has value $\log k_{J,K}^{pot} a_K(1)$. Thus in the further discussion the operational value of the selectivity coefficient will be determined by setting $a_J = 0$, converting all the a_K functions into a single logarithmic function and dividing the argument by a_K or a certain suitable power of a_K.

In some types of ISE, the inner side of the membrane is in immediate contact with a metal phase. The properties of these ISEs will be discussed in chapter 4.

This chapter is concerned with processes that lead to formation of ISE membrane potentials. The membrane potentials of electrodes with liquid membranes containing a dissolved ion-exchanger ion or a dissolved ionophore (ion carrier), and of electrodes with solid or glassy membranes will be considered. More complicated systems, for example ISEs with a gas gap and enzyme electrodes, will be discussed in chapters 4 and 9.

This chapter is based on the thermodynamic theory of membrane potentials and kinetic effects will be considered only in relation to diffusion potentials in the membrane. The ISE membrane in the presence of an interferent can be thought of as analogous to a corroding electrode [46a] at which chemically different charge transfer reactions proceed [15, 16]. Then the characteristics of the ISE potentials can be obtained using polarization curves for electrolysis at the boundary between two immiscible electrolyte solutions [44a, 46].

All ISE membranes are compact phases through which ions and the solvent from the solution with which the membrane is in contact cannot freely pass. These phases may be homogeneous or heterogeneous, for example a polymeric structure containing tiny interconnected droplets of organic liquid in which ion-exchanger ions are dissolved (termed solvent–polymeric membranes, see p. 61). On the other hand, hydrophilic porous membranes of polymeric materials with fixed ions of a single charge on the pore walls (ion-exchanger membranes) have a number of technical applications (electrodialysis, membrane electrolysis, etc.), but have not been found to be useful as ISE membranes. The theory of this type of electrochemical membranes [52, 53, 76, 77] is described in [44].

It should be borne in mind that theoretical considerations of membrane potentials are based on a number of assumptions about quantities that are not known or cannot even be found. This is especially true of the ion mobility in a solid or liquid membrane phase, of the dependence of the mobility in the liquid phase on the concentration, of the activities of ions in the membrane phases, etc. The subsequent discussion will deal only with those parts of the theory needed to understand the basic properties of the systems as they appear during practical measurements. Refined theories are accompanied by so many assumptions that they cannot be used for precise predictions of the behaviour of such systems.

3.2 Membranes with dissolved ion-exchanger ions

These are discussed in [2, 4, 11, 12, 13, 17, 18, 19, 20, 22, 23, 24, 34, 35, 36, 37, 50, 51, 54, 65, 66, 67, 68, 69, 71, 79]. This type of liquid membrane contains strongly hydrophobic ions (ion-exchanger ions) and the determinand ion with the opposite charge (counter-ion, gegenion). Their behaviour depends on the distribution coefficient of the salt of the ion-exchanger ion with the determinand between the aqueous solution and the membrane solvent, on the formation of ion pairs in the membrane and on the degree to which interfering ions can enter the membrane. Also important is whether the charge numbers of the interfering ion and of the determinand are the same or different, whether a diffusion potential is formed in the membrane in the presence of several kinds of counter-ion, and finally, the degree to which hydrophobic co-ions (ions with the same charge sign as that of the ion-exchanger ion) enter the membrane. All these factors will be discussed here.

First consider the system in which no diffusion potential is formed in the membrane. The membrane potential is then determined by the conditions at the membrane/aqueous electrolyte solution boundary. In the simplest situation, a salt of a monovalent ion-exchanger ion, anion A^-, with monovalent determinand cation J^+ is dissolved in the membrane. In order for this system to be the basis for a usable ISE with Nernstian response to the determinand ion in a sufficiently broad activity interval, it is necessary that the distribution coefficient k_{JA} be

$$k_{JA} = \exp\left(-G^{0,\,wt\to m}_{tr,\,JA}/RT\right) \gg 1 \qquad (3.2.1)$$

where $\Delta G^{0,\,wt\to m}_{JA}$ is the standard Gibbs energy for transfer of salt JA from water (wt) into the membrane (m) (see p. 12).

As ion-exchanger anion A^- is strongly hydrophobic and determinand J^+ is hydrophilic, it must hold for the individual Gibbs energies of transfer for the two ions that $\Delta G^{0,\,wt\to m}_{tr,\,A^-} \ll \Delta G^{0,\,wt\to m}_{tr,\,J^+}$.

The potential difference at the phase boundary between the membrane and solution 1 (cf. (2.2.11)) is

$$\Delta^m_1 \phi = -\Delta^{wt}_m \phi^0_{J^+} + (RT/F) \ln\left[a_{J^+}(1)/a_{J^+}(m)\right], \qquad (3.2.2)$$

where $a_{J^+}(m)$ and $a_{J^+}(1)$ are the activities of ion J^+ in the membrane and in solution 1. Because an analogous relationship also holds for the membrane/solution 2 boundary, the overall membrane potential (cf. (3.1.4) is

$$\Delta\phi_M = \phi(2) - \phi(1) = (RT/F) \ln [a_{J^+}(1)/a_{J^+}(2)]. \tag{3.2.3}$$

It apparently holds for determinand J^{z+} that

$$\Delta\phi_M = (RT/zF) \ln [a_{Jz^+}(1)/a_{Jz^+}(2)]. \tag{3.2.4}$$

The ISE potential can then be obtained from (3.1.5).

The usefulness of the ion-selective electrode for determining ion J^+ is limited by the distribution equilibrium for salt JA between the membrane and solution 1. If the activity of ion J^+, which is present in solution 1 only as a result of this distribution equilibrium, is much smaller than the overall activity $a_{J^+}(1)$, then the dependence of the membrane potential on the activity of determinand J^+ has Nernstian slope. If this condition is not fulfilled, then the ISE behaviour can be obtained from the equation for the distribution equilibrium

$$k_{JA}^2 = [a_{J^+}(m)a_{A^-}(m)/a_{J^+}(1)a_{A^-}(1)]. \tag{3.2.5}$$

It will be assumed that, on transfer of a certain amount of salt JA from the membrane into solution 1, activities $a_{J^+}(m)$ and $a_{A^-}(m)$ practically do not change. Activity $a_{J^+}(1)$ depends both on the original concentration of ion J^+ in the studied solution, $c_{J^+}^0(1)$ and on the concentration corresponding to the ions leaving the membrane $c_{J^+}'(1)$,

$$a_{J^+}(1) = \gamma_{J^+}[c_{J^+}^0(1) + c_{J^+}'(1)]. \tag{3.2.6}$$

The activity of the anion $a_{A^-}(1)$ depends on the ion concentration $c_{J^+}'(1)$

$$a_{A^-}(1) = \gamma_{A^-}c_{J^+}'(1). \tag{3.2.7}$$

Substitution of (3.2.6) and (3.2.7) into (3.2.5) and rearrangement yields

$$\Delta\phi_M = \frac{RT}{F} \ln \left(\frac{a_{J^+}^0(1) + \{a_{J^+}^0(1)^2 + [4a_{\pm,JA}(m)^2/\gamma_{A^-}k_{JA}^2]\}^{1/2}}{2a_{J^+}(2)} \right)$$

$$\tag{3.2.8}$$

where $a_{J^+}^0(1)$ is the activity of the determinand in the bulk of the analyte and $a_{\pm,JA}(m)$ is the mean activity of salt JA in the membrane. It should be noted that changes in concentrations J^+ and A^- are observed only in the vicinity of the membrane/studied solution boundary.

If the first term in the sum under the square root sign is much larger than the second term, then (3.2.8) reverts to (3.2.3). In the opposite case, the ISE potential is no longer dependent on the determinand activity because then

$$\Delta\phi_M = \frac{RT}{F} \ln \frac{a_{\pm,JA}(m)}{\gamma_A^{1/2} k_{JA} a_{J^+}(2)}. \tag{3.2.9}$$

This equation characterizes the detection limit for determining the determinand activity, as depicted in fig. 3.1. Curve 1 of fig. 3.1 is plotted from the Nikolsky equation (3.1.7) for the product $k_{J,K}^{pot} a_{K^+}(1) = 10^{-4}$, $z = 1$ and const $= 0$. Curve 2 gives the limit of detection according to the equation (cf. 3.2.8)

$$E_{ISE} = const + (RT/F) \ln (\tfrac{1}{2} \{a_{J^+}^0(1)$$
$$+ [a_{J^+}^0(1)^2 + (4a_{\pm, JA}(m)^2/\gamma_A \cdot k_{JA}^2)]^{1/2}\})$$

for $(4a_{\pm, JA}(m)^2/\gamma_A \cdot k_{JA}^2) = 10^{-10}$ and const $= -20$ mV.

If the ion-exchanger ion forms an ion pair in the membrane, characterized by the ion-pair formation constant

$$K_{JA} = \frac{a_{JA}}{a_{J^+} a_{A^-}}, \tag{3.2.10}$$

then (3.2.3) and (3.2.4) remain valid. Activity $a_{J^+}(m)$ is affected by ion pair formation, so that when association predominates it holds approximately that

$$a_{J^+}(m) = \left(\frac{c_{JA}}{K_{JA}}\right)^{1/2}, \tag{3.2.11}$$

where c_{JA} is the concentration of salt JA in the membrane. Association leads to a decrease in the detection limit because (3.2.9) is converted into the approximate form

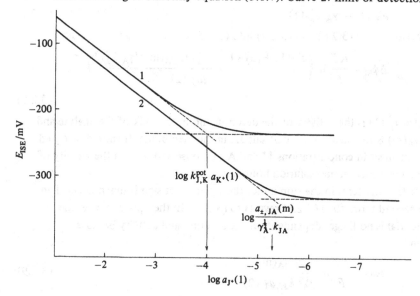

Fig. 3.1. Dependence of E_{ISE} on activity of the determinand J^+. Curve 1: data according to Nikolsky equation (3.1.7). Curve 2: limit of detection.

$$\Delta\phi_M \approx \frac{RT}{F}\ln\left(\frac{c_{JA}^{1/2}}{k_{JA}K_{JA}^{1/2}a_{J^+}(2)}\right). \tag{3.2.12}$$

Consider a test solution containing both determinand J^+ and interferent K^+, neither of which forms an ion-pair with ion-exchanger ion A^-. The mobilities of J^+ and K^+ are identical, so that no diffusion potential is formed in the membrane. The effect of the interferent is based on the fact that it may replace the determinand in the membrane phase as a result of the exchange reaction

$$J^+(m) + K^+(1) = J^+(1) + K^+(m). \tag{3.2.13}$$

For the equilibrium constant of this reaction,

$$K_{exch} = \frac{a_{J^+}(1)a_{K^+}(m)}{a_{J^+}(m)a_{K^+}(1)} = \exp\left(\frac{\Delta G_{tr,J^+}^{0,\,wt\to m} - \Delta G_{tr,K^+}^{0,\,wt\to m}}{RT}\right) \tag{3.2.14}$$

If determinand J^+ is displaced from the membrane (at least in the region around the membrane/solution 1 boundary), then (3.2.13) indicates that an increase in the activity of the determinand in the test solution increases its activity in the membrane in the same proportion, so that the ratio does not change (see (3.2.2)) and the potential difference at the membrane/solution 1 boundary also does not change. This fact can be expressed quantitatively using (3.2.14). For the concentrations of ions J^+ and K^+ in the membrane,

$$c_{J^+}(m) + c_{K^+}(m) = \frac{a_{J^+}(m)}{\gamma_{J^+}} + \frac{a_{K^+}(m)}{\gamma_{K^+}} = c^0, \tag{3.2.15}$$

where c^0 is the concentration of ion-exchanger in the membrane.

(3.2.14) and (3.2.15) yield the concentrations of ions J^+ and K^+ in the membrane:

$$c_{J^+}(m) = \frac{c^0 a_{J^+}(1)}{a_{J^+}(1) + (\gamma_{J^+}/\gamma_{K^+})K_{exch}a_{K^+}(1)}$$

$$c_{K^+}(m) = \frac{(\gamma_{J^+}/\gamma_{K^+})K_{exch}c^0 a_{K^+}(1)}{a_{J^+}(1) + (\gamma_{J^+}/\gamma_{K^+})K_{exch}a_{K^+}(1)} \tag{3.2.16}$$

Substitution of (3.2.15) into (3.2.14) and subsequent substitution into (3.2.2) yields

$$\Delta_1^m\phi = -\Delta_m^{wt}\phi_{J^+}^0 + \frac{RT}{F}\ln\frac{a_{J^+}(1) + (\gamma_{J^+}/\gamma_{K^+})K_{exch}a_{K^+}(1)}{\gamma_{J^+}c^0} \tag{3.2.17}$$

where γ_{J^+} and γ_{K^+} are the activity coefficients in the membrane phase and c^0 is the ion-exchanger ion concentration in the membrane. As the internal ISE solution contains only the determinand, the membrane potential is

$$\Delta\phi_{M} = \frac{RT}{F} \ln\left[\frac{a_{J^+}(1) + (\gamma_{J^+}/\gamma_{K^+})K_{exch}a_{K^+}(1)}{a_{J^+}(2)}\right].$$ (3.2.18)

The ISE potential is given by (3.1.7), where it holds for the selectivity coefficient that

$$k^{pot}_{J^+, K^+} = (\gamma_{J^+}/\gamma_{K^+})K_{exch}.$$ (3.2.19)

Fig. 3.2 shows the dependence of the ISE potential on the logarithm of the activity of ion J^+ for various values of the activity of ion K^+ (calculated for $k^{pot}_{J, K} = 0.1$ and const $= 0$).

Consider a system in which the membrane ions form stable ion-pairs with both ions, whose equilibrium formation constants are K_{JA} and K_{KA}. Then, according to (3.2.10), (3.2.14) converts to

$$K_{exch}K_{KA}/K_{JA} = a_{J^+}(1)a_{KA}/a_{K^+}(1)a_{JA}.$$ (3.2.20)

As most of the J^+ and K^+ ions are ion-paired, then $c_{JA} + c_{KA} = c_0$, where c_0 is the overall concentration of ion-exchanger ion A^-. The same procedure as used above yields the equations for the membrane potential and the ISE potential.

Fig. 3.2. Dependence of E_{ISE} on the activity of the determinand J^+ at various activities of the inerferent K^+ (3.1.7).

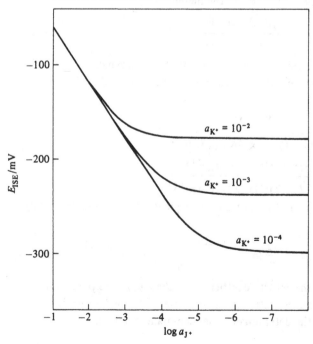

It follows for the selectivity coefficient that

$$k_{J^+,K^+}^{pot} = K_{exch} K_{KA}/K_{JA}. \tag{3.2.21}$$

The ratio of the activity coefficients of ion pairs JA and KA was assumed equal to one.

A further system is characterized by the determinand and interferent having *different charge numbers*, for example $+1$ and $+2$. Neither ion pairs nor diffusion potential are formed in the membrane. The exchange reaction is then

$$2J^+(m) + K^{2+}(1) = 2J^+(1) + K^{2+}(m), \tag{3.2.22}$$

with ion exchange equilibrium constant

$$K_{exch} = a_{J^+}(1)^2 a_{K^{2+}}(m)/a_{J^+}(m)^2 a_{K^{2+}}(1). \tag{3.2.23}$$

For the concentrations of both ions in the membrane,

$$c_{J^+} + 2c_{K^{2+}} = c_{A^-}. \tag{3.2.24}$$

Substitution into (3.2.22) and then into (3.2.2) yields the potential at the membrane/solution 1 boundary:

$$\Delta_1^m \phi = -\Delta_m^{wt} \phi_{J^+}^0$$
$$+ \frac{RT}{F} \ln \frac{a_{J^+}(1) + [a_{J^+}(1)^2 + 8K_{exch} a_{K^{2+}}(1) c_A \gamma_{J^+}^2 \gamma_{K^{2+}}^{-1}]^{1/2}}{2\gamma_{K^{2+}} c_0}$$

$$\tag{3.2.25}$$

where γ_{J^+} and $\gamma_{K^{2+}}$ are the activity coefficients of J^+ and K^{2+} in the membrane phase. It follows for the ISE potential that

$$E_{ISE} = \text{const} + \frac{2.3\,RT}{F}$$
$$\times \log\left(\tfrac{1}{2}\{a_{J^+}(1) + [a_{J^+}(1)^2 + 8K_{exch}\gamma_{J^+}^2 \gamma_{K^{2+}}^{-1} c_A a_{K^{2+}}(1)]^{1/2}\}\right)$$

$$\tag{3.2.26}$$

It can be seen that different charge numbers on the participating ions considerably complicate the resultant expression, which no longer has the form of the Nikolsky equation (3.1.7) (similar complications occur in the Planck liquid–junction theory). The E_{ISE} versus $\log a_{J^+}(1)$ dependence (see fig. 3.3, plotted for $K_{exch}\gamma_{J^+}^2 \gamma_{K^{2+}} c_A = 1.25 \times 10^{-11}$ and const $= 0$) is characterized by asymptotes

$$E_{ISE} = \text{const} + 2.3RT/F \log [a_{J^+}(1)],$$

$$E_{ISE} = \text{const} + 2.3RT/2F \log [8K_{exch}\gamma_{J^+}^2 \gamma_{K^{2+}}^{-1} c_A a_{K^{2+}}(1)],$$

intersecting at point

$$\log a_{J^+}(1) = \tfrac{1}{2} \log [2K_{exch}\gamma_{J^+}^2 \gamma_{K^{2+}}^{-1} c_A a_{K^{2+}}(1)].$$

Quantity $(2K_{exch}\gamma_{J^+}^2 \gamma_{K^{2+}}^{-1} c_A)^{1/2}$ can again be taken as the selectivity coefficient (see p. 28). The shapes of the dependence according to (3.2.26) and according to the modified Nikolsky equation for determinand J^+ and interferent K^{2+}

$$E_{ISE} = \text{const} + (2.3\,RT/F)$$
$$\times \log\,[a_{J^+}(1) + k^{pot}_{J^+,K^{2+}}a_{K^{2+}}(1)^{1/2}] \tag{3.2.27}$$

are not very different (see fig. 3.3) as the asymptotes to this dependence expressed in the E_{ISE} versus $\log a_{J^+}(1)$ coordinates intersect at the same point. The selectivity coefficient is then directly proportional to the ion-exchanger ion concentration on the membrane.

In addition to interferents with the same charge sign, ions with the opposite sign can also interfere if they are present in the test solution in an appropriate concentration and if they have sufficiently low standard Gibbs free energy of transfer from water into the membrane solvent. Consider an aqueous solution containing determinand J^+, a corresponding hydrophilic anion and a relatively hydrophobic anion B^-. The presence of the latter anion in the test solution leads to transfer of salt JB into the membrane phase, so that the membrane potential depends on the concentration of this anion and no longer obeys the Nernstian dependence on the activity of ion J^+.

Fig. 3.3. Dependence of E_{ISE} on the activity of a monovalent determinand J^+ in the presence of a divalent interferent K^{2+} (equation (3.2.26), full line) compared with the Nikolsky equation (equation (3.2.27), dotted line).

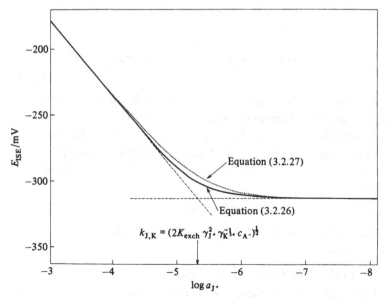

Equation (3.2.5) can again be used, where the indices referring to ion A^- are replaced by those for anion B^-. The passage of a certain amount of salt JB into the membrane phase leads to an increase in the concentration of ion J^+ over the original value c_0 to value c_{J^+}. The concentration of ion B^- in the membrane will be $c_{J^+} - c_0$. The distribution equilibrium equation is then converted to

$$k_{JB}^2 = \frac{\gamma_{J^+}\gamma_{B^-}c_{J^+}(c_{J^+} - c_0)}{a_{J^+}(1)a_{B^-}(1)}. \tag{3.2.28}$$

Substitution of this equation into (3.2.2) yields

$$\Delta_1^m \phi = -\Delta_m^{wt}\phi_{J^+}^0$$
$$+ \frac{RT}{F} \ln \frac{2a_{J^+}(1)}{\gamma_{J^+}c_0 \{1 + [1 + (4k_{JB}^2 a_{J^+}(1)a_{B^-}(1)/\gamma_{J^+}\gamma_{B^-}c_0^2)]^{1/2}\}} \tag{3.2.29}$$

It is apparent that the deviation from Nernstian behaviour depends on the activity of the determinand and anion B^- in the studied solution. It decreases with increasing magnitude of the sum of the standard Gibbs energies of transfer of ions J^+ and B^- from water into the membrane phase. The effect of the interfering anion is suppressed by increasing the concentration of the ion-exchanger ion in the membrane.

Consider a system in which the analyte contains both determinand J^+ and interferent K^+, and where a diffusion potential is formed in the membrane as a result of their different mobilities. A simplification that provides the basic characteristics of the membrane potential employs the Henderson equation for calculation of the diffusion potential in the membrane. According to (2.1.9) the membrane potential is separated into three parts, two potential differences between the membrane and the solutions $\Delta_1^p\phi$ and $\Delta_q^2\phi$ with which it is in contact, and the diffusion potential inside the membrane

$$\Delta\phi_M = \phi(2) - \phi(1)$$
$$= \phi(2) - \phi(q) + \phi(q) - \phi(p) + \phi(p) - \phi(1)$$
$$= \Delta_q^2\phi + \Delta\phi_L + \Delta_1^p\phi, \tag{3.2.30}$$

where p and q designate the coordinates of the inner sides of the membrane. Potential differences $\Delta_q^2\phi$ and $\Delta_1^p\phi$ are given by (3.2.17), which simplifies for $\Delta_q^2\phi$ to

$$\Delta_q^2\phi = \Delta_m^{wt}\phi_{J^+}^0 + \frac{RT}{F} \ln \frac{\gamma_{J^+}c_0}{a_{J^+}(2)}. \tag{3.2.31}$$

The concentrations of ions J^+ and K^+ at point p, which are needed to calculate the diffusion potential, are given by (3.2.16), while the concentration of ion J^+ at point q is c_0. It follows from (2.6.12) that

$$\Delta\phi_L = -\frac{RT}{F} \ln \frac{(U_{J^+}/U_{K^+})c_0}{(U_{J^+}/U_{K^+})c_{J^+}(m) + c_{K^+}(m)}, \tag{3.2.32}$$

where U_{J^+} and U_{K^+} are the mobilities of ions J^+ and K^+, respectively, and $c_{J^+}(m)$ and $c_{K^+}(m)$ are given by relationships (3.2.16). Substitution of (3.2.17), (3.2.31) and (3.2.32) into (3.2.30) yields the relationship for the membrane potential

$$\Delta\phi_M = \frac{RT}{F} \ln \frac{a_{J^+}(1) + (\gamma_{J^+}/\gamma_{K^+})(U_{K^+}/U_{J^+})K_{exch}a_{K^+}(1)}{a_{J^+}(2)}. \tag{3.2.33}$$

Thus the Nikolsky equation (3.1.7) with selectivity coefficient

$$k^{pot}_{J^+, K^+} = \frac{\gamma_{J^+}}{\gamma_{K^+}} \frac{U_{K^+}}{U_{J^+}} K_{exch} \tag{3.2.34}$$

is valid for the ISE potential in the presence of both the determinand and interferent in the analyte. It is also valid when there is a diffusion potential in the membrane, and when there is negligible ion association.

Finally, the result of a theoretical treatment of a similar system with almost complete association in the membrane will be given without calculations [66, 67]. The diffusion potential in the membrane depends not only on electrodiffusion of J^+, K^+ and A^- but also on diffusion of associates JA and KA. The resultant formula for the membrane potential is

$$\Delta\phi_M = \frac{RT}{F}(1-\tau)\ln$$
$$\times \frac{(U_{J^+} + U_{A^-})a_{J^+}(1) + (U_{K^+} + U_{A^-})K_{exch}a_{K^+}(1)}{(U_{J^+} + U_{A^-})a_{J^+}(2)}$$
$$+ \frac{RT}{F}\tau\ln \frac{U_{JA}K_{JA}a_{J^+}(1) + U_{KA}K_{KA}K_{exch}a_{K^+}(1)}{U_{JA}K_{JA}a_{J^+}(2)}, \tag{3.2.35}$$

where the transport number τ is given by

$$\tau = \frac{U_{A^-}(U_{KA}K_{KA} - U_{JA}K_{JA})}{(U_{J^+} + U_{A^-})U_{KA}K_{KA} - (U_{K^+} + U_{A^-})U_{JA}K_{JA}}. \tag{3.2.36}$$

If the ion-exchanger ion A^- is far more mobile than free ions J^+ and K^+ in the membrane, then the second term in (3.2.35) can be neglected and the membrane potential is

$$\Delta\phi_M = \frac{RT}{F} \ln \frac{a_{J^+}(1) + (U_{KA}K_{KA}/U_{JA}K_{JA})a_{K^+}(1)}{a_{J^+}(2)} \tag{3.2.37}$$

so that the selectivity coefficient has the value

$$k^{pot}_{J^+,K^+} = \frac{U_{KA}}{U_{JA}} \frac{K_{KA}}{K_{JA}}. \tag{3.2.38}$$

In general, however, the resultant dependence has three asymptotic parts with intercepts at

$$a_{J^+}(1) = \frac{(U_{K^+} + U_{A^-})K_{exch}}{U_{J^+} + U_{A^-}} a_{K^+}(1),$$

$$a_{J^+}(1) = \frac{U_{KA}K_{KA}K_{exch}}{U_{JA}K_{JA}} a_{K^+}(1). \tag{3.2.39}$$

The selectivity coefficient obtained by the procedure on p. 35 is given in general by

$$k^{pot}_{J^+,K^+} = \left(\frac{U_{K^+} + U_{A^-}}{U_{J^+} + U_{A^-}}\right)^{1-\tau} \left(\frac{U_{KA}K_{KA}}{U_{JA}K_{JA}}\right)^{\tau} K_{exch}. \tag{3.2.40}$$

This theory was developed further by Sandblom and coworkers [68] for a system where 'side reactions' occur in the membrane, such as adduct formation and polymerization.

In conclusion, it should be mentioned that extraction parameters (the equilibrium constants of exchange reactions and ion-pair stabilities) were introduced into the theory of ion-selective electrodes in [2, 31, 33, 34, 35, 69]. The theory of ISEs with a liquid membrane and a diffusion potential in the membrane was extended by Buck *et al.* [11, 13, 14, 73, 74] and Morf [54]. The theory of the effect of co-ions (an ion with the same charge sign) which enter the membrane (see (3.2.29)) was developed by Jyo and coworkers [36] and by Stover and Buck [74].

3.3   Membranes with a neutral ion carrier (ionophore)

These liquid membranes contain a dissolved complex of the determinand with a strongly hydrophobic complexing agent, whose name, ionophore, is taken from membrane biology [61]. In the past, a great deal of effort has been devoted to developing a theory of the original forms of these systems, in which conditions are complicated by the extraction of a poorly defined amount of determinand together with a hydrophilic counterion [6, 17, 25, 26, 42, 47, 55, 60, 64, 72, 75, 78]. A considerable step forward was made by Morf *et al.* [55], who developed a system with definite membrane composition and simple behaviour towards the test solution. The discussion below is concerned only with this system, because it has clear advantages, for practical ISEs, over other systems.

Consider a liquid membrane in which a complex of determinand J^+ and complexing agent (ionophore) X is dissolved. In the complex a single ion J^+ is bonded to a single X molecule. The complex is so strong that the concentration of free ions is negligible compared to the concentration of ions bound in the complex. The concentration of the determinand in the membrane is fixed by the concentration of very hydrophobic anions A^-, whose concentration in aqueous solutions in contact with the membrane is negligible. Similarly, the concentration of the ionophore in water is negligible because of its strong hydrophobicity. Consider two practically important types of membrane system:

1. the membrane contains only the complex without excess ionophore;
2. the membrane contains both the complex and excess ionophore.

In the first system, the membrane/test solution potential difference is given by (3.2.2). For equilibrium in the membrane,

$$K_{JX} = \frac{a_{JX^+}(m)}{a_{J^+}(m)a_X(m)} = \frac{\gamma_{JX^+}\gamma_{J^+}}{\gamma_X}\frac{c_0}{a_{J^+}(m)^2}, \tag{3.3.1}$$

where K_{JX} is the stability constant of complex JX^+ in the membrane phase and c_0 is the concentration of the salt of the ionophore complex with ion J^+ and anion A^-. If $a_{J^+}(m)$ is calculated from (3.3.1) and substituted into (3.2.2), then an equation is obtained for the membrane/solution 1 potential:

$$\Delta_1^m\phi = -\Delta_m^{wt}\phi_{J^+}^0 + \frac{RT}{F}\ln a_{J^+}(1)$$

$$- \frac{RT}{2F}\ln\left(\frac{\gamma_{J^+}\gamma_{JX^+}}{\gamma_X}\frac{c_0}{K_{JX}}\right). \tag{3.3.2}$$

Because a similar equation holds for the membrane/solution 2 phase boundary and no diffusion potential is formed within the membrane, (3.2.3) is valid for the membrane potential.

The formation of ion-pairs between complex JX^+ and anion A^- in the membrane, provided that the membrane solvent is not very polar, does not affect the Nernstian response of the ISE and appears only in the constant term in (3.3.2). If most of the complex is in the form of ion-pair JXA but the concentration of ion-pair JA is much less than the concentration of JXA, then for the activity of the free complex ions

$$a_{JX^+} = \left(\frac{\gamma_{J^+}\gamma_{JXA}}{\gamma_{A^-}}\frac{c_0}{K_{JXA}}\right)^{1/2}, \tag{3.3.3}$$

where K_{JXA} is the formation constant for ion-pair JXA. Substitution into (3.3.1) and then into (3.2.2) yields the relationship for $\Delta_1^m\phi$:

$$\Delta_1^m\phi = -\Delta_m^{wt}\phi_{J^+}^0 + \frac{RT}{F}\ln a_{J^+}(1)$$

$$-\frac{RT}{4F} \ln \left(\frac{\gamma_{J^+}^2 \gamma_{JX^+} \gamma_{JXA}}{\gamma_X^2 \gamma_{A^-}} \frac{c_0}{K_{JX}^2 K_{JXA}} \right) \tag{3.3.4}$$

The membrane potential is given by (3.2.3).

The formation of ion pairs in the membrane thus has no effect on the Nernstian response of the membrane.

If the test solution contains at the same time determinand J^+ and interferent K^+, then for the overall complex concentration in the membrane

$$c_0 = \frac{a_{JX^+}(m)}{\gamma_{JX^+}} + \frac{a_{KX^+}(m)}{\gamma_{KX^+}} . \tag{3.3.5}$$

Also,

$$a_X = \gamma_X \left(\frac{a_{J^+}(m)}{\gamma_{J^+}} + \frac{a_{K^+}(m)}{\gamma_{K^+}} \right). \tag{3.3.6}$$

Substitution of the equation for the stability of the complexes, (3.3.1) yields

$$\begin{aligned}
c_0 &= \frac{K_{JX} a_X(m) a_{J^+}(m)}{\gamma_{JX^+}} + \frac{K_{KX} a_X(m) a_{K^+}(m)}{\gamma_{KX^+}} \\
&= a_{J^+}(m)^2 K_{JX} \frac{\gamma_X}{\gamma_{J^+} \gamma_{JX^+}} \left(1 + \frac{K_{KX}}{K_{JX}} \frac{\gamma_{JX^+}}{\gamma_{KX^+}} \frac{a_{K^+}(m)}{a_{J^+}(m)} \right) \\
&\times \left(1 + \frac{\gamma_{J^+}}{\gamma_{K^+}} \frac{a_{K^+}(m)}{a_{J^+}(m)} \right).
\end{aligned} \tag{3.3.7}$$

Introduction of relationship (3.2.14) yields

$$\begin{aligned}
a_{K^+}(m) &= a_{J^+}(m) \exp \left(\frac{\Delta G_{tr, J^+}^{0, wt \to m} - G_{tr, K^+}^{0, wt \to m}}{RT} \right) \frac{a_{K^+}(1)}{a_{J^+}(1)} \\
&= \frac{a_{J^+}(m) K_{exch} a_{K^+}(1)}{a_{J^+}(1)} .
\end{aligned} \tag{3.3.8}$$

Substitution into (3.3.6) and then into (3.2.2) leads to

$$\begin{aligned}
\Delta_1^m \phi &= -\Delta_m^{wt} \phi_{J^+}^0 + \frac{RT}{2F} \ln \left[a_{J^+}(1) \right. \\
&\quad \left. + \frac{K_{KX}}{K_{JX}} \frac{\gamma_{JX^+}}{\gamma_{KX^+}} K_{exch} a_{K^+}(1) \right] \\
&\quad + \frac{RT}{2F} \ln \left[a_{J^+}(1) + \frac{\gamma_{J^+}}{\gamma_{K^+}} K_{exch} a_{K^+}(1) \right] \\
&\quad + \frac{RT}{2F} \ln \left(\frac{K_{JX}}{c_0} \frac{\gamma_X}{\gamma_{J^+} \gamma_{JX^+}} \right).
\end{aligned} \tag{3.3.9}$$

As the complexes of various ions with a single ionophore usually have the same structure (they are 'isosteric' [17, 23, 25]), their mobilities in the membrane are the same and consequently no diffusion potential is formed in the membrane. The selectivity coefficient is then (see p. 35):

$$k_{J^+,K^+}^{pot} = \left(\frac{K_{KX}}{K_{JX}} \frac{\gamma_{J^+}}{\gamma_{K^+}} \frac{\gamma_{JX^+}}{\gamma_{KX^+}} \right)^{1/2} K_{exch}. \tag{3.3.10}$$

Introduction of ionophores into the membrane should lead to a considerable increase in the ISE selectivity (in the absence of an ionophore, the selectivity coefficient would be given by constant K_{exch} alone, cf. (3.3.12)); however, as (3.3.10) contains only the square root of the stability constant ratio, this increase in selectivity is not extremely high. Better results are obtained in the second system, where the membrane contains an excess constant ionophore concentration.

Under these conditions it follows from the first part of (3.3.1) that

$$a_{J^+}(m) = \frac{\gamma_{JX^+} c_0}{K_{JX} \gamma_X c_X}, \tag{3.3.11}$$

where c_0 is the concentration of the complex in the membrane, equal to the concentration of hydrophobic anions and c_X is the concentration of excess free ionophore. The membrane potential is apparently given by (3.2.3). If a stable ion pair JXA is formed in the membrane, then the activity of the complex is given by (3.3.3) and the activity of free ion J^+ in the membrane by

$$a_{J^+} = \frac{1}{K_{JX^+} \gamma_X c_X} \left(\frac{\gamma_{J^+} \gamma_{JXA} c_0}{\gamma_{A^-} K_{JXA}} \right)^{1/2}. \tag{3.3.12}$$

The membrane potential is again given by (3.2.3). Limitation of the Nernstian response occurs in both systems if the analyte contains a salt of J^+ with rather hydrophobic anion B^-, which can dissolve in the membrane to form a further complex.

The effect of anion B^- on the potential of the ISE was expressed by Morf *et al.* [57] by the relationship

$$E_{ISE} = const - (RT/F) \ln [a_{B^-}(1) + k_{B^-,J^+}^{pot} a_{J^+}^{-1}(1)]. \tag{3.3.13}$$

This equation contains selectivity coefficient k_{B^-,J^+}^{pot} for cation J^+ with respect to anion B^-; consequently, activity $a_{J^+}(1)$ has a negative exponent. The behaviour of the ISE in fig. 3.4 can be explained in this manner.

If the test solution contains both determinand J^+ and interferent K^+, their activities in the membrane are given by (3.3.5), the first of (3.3.7) and (3.3.8). Consequently,

$$\Delta_1^m \phi = -\Delta_m^{wt}\phi_{J^+}^0 + \frac{RT}{F}\ln$$

$$\times \left[a_{J^+}(1) + \frac{K_{KX}}{K_{JX}}\frac{\gamma_{JX}}{\gamma_{KX}}K_{exch}a_{K^+}(1)\right] + \frac{RT}{F}\ln\frac{\gamma_X c_X K_{JX}}{\gamma_{JX^+}c_0}.$$

(3.3.14)

Assuming that the mobilities of both complexes are practically the same because of their structures [17, 23, 25] so that the diffusion potential in the membrane is thus negligible, this equation readily yields relationships for the membrane and ISE potentials.

The selectivity coefficient is given by

$$k_{J^+,K^+}^{pot} = \frac{K_{KX}}{K_{JX}}\frac{\gamma_{JX^+}}{\gamma_{KX^+}}K_{exch}.$$

(3.3.15)

Fig. 3.4. Hydrophobic anion effect on E_{ISE}. The membrane contains valinomycin with equivalent concentration of potassium tetraphenyl-borate dissolved in 2-nitro-p-cymene. △ - E_{ISE} dependence on the activity of KCl; ○ - E_{ISE} dependence on the activity of KClO$_4$. The curves were calculated using equation (3.3.13). (After W. E. Morf, G. Kahr and W. Simon.)

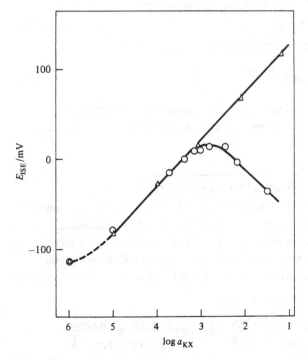

In the literature, the selectivity coefficient value is usually given as $k_{J^+,K^+}^{pot} = K'_{KX}/K'_{JX}$, where K'_{KX} and K'_{JX} are the stability constants of the complexes in the aqueous phase. This value is identical to the value from (3.3.15) because

$$K'_{KX}/K'_{JX} = [a_{KX^+}(wt)/a_{K^+}(wt)a_X(wt)]/[a_{JX^+}(wt)/a_{J^+}(wt)a_X(wt)]$$
$$= \{[a_{KX^+}(m)/a_{K^+}(m)]/[a_{JX^+}(m)/a_{J^+}(m)]\}$$
$$\times [k_{K^+}k_{JX^+}/k_{J^+}k_{KX^+}] \approx K_{exch}(K_{KX}/K_{JX}),$$

where

$$k_i = \exp(-\Delta G_{tr,i}^{0,\,wt \to m})$$

is the individual distribution coefficient for substance i between the membrane solvent and water. As the complexes have almost identical structures, it is assumed that $k_{JX^+} \approx k_{KX^+}$. Because determination of the stability constant of the ionophore complexes in the organic phase and of the standard Gibbs energies are experimentally relatively simple, expression of the selectivity coefficient according to (3.3.15) is preferred here.

Some synthetic ionophores form complexes with metallic ions containing a varying number of ligands per metal ion, usually 1 or 2. It then holds for the overall concentration of ion J^+ in the membrane that

$$c_0 = c_{J^+} + c_{JX^+} + c_{JX_2^+}. \qquad (3.3.16)$$

If c_{J^+} is neglected and the stability constants

$$\beta_{JX^+} = \frac{a_{JX^+}}{a_{J^+}a_X}, \quad \beta_{JX_2^+} = \frac{a_{JX_2^+}}{a_{J^+}a_X^2} \qquad (3.3.17)$$

are employed to express c_{JX^+} and $c_{JX_2^+}$, then a relationship is obtained for $a_{J^+}(m)$ (it is assumed that ionophore X is in excess and that its concentration is constant):

$$a_{J^+}(m) = \frac{c_0}{(\beta_{JX^+}c_X/\gamma_{JX^+}\gamma_X) + (\beta_{JX_2^+}c_X^2/\gamma_{JX_2^+}\gamma_X^2)}. \qquad (3.3.18)$$

Substitution into (3.2.2) yields a simple relationship for the membrane/solution potential and thus also expressions for the membrane and ISE potentials.

If the test solution contains determinand J^+ and interferent K^+, it follows for their overall concentration (equal to the concentration of the hydrophobic anion, neglecting the concentrations of free J^+ and K^+ ions in the membrane) that

$$c_0 = c_{JX^+} + c_{JX_2^+} + c_{KX^+} + c_{KX_2^+}$$
$$= a_{J^+}(m)\left[\frac{\beta_{JX^+}c_X}{\gamma_{JX^+}\gamma_X} + \frac{\beta_{JX_2^+}c_X^2}{\gamma_{JX_2^+}\gamma_X^2}\right] + a_{K^+}(m)\left[\frac{\beta_{KX^+}c_X}{\gamma_{KX^+}\gamma_X} + \frac{\beta_{KX_2^+}c_X^2}{\gamma_{KX_2^+}\gamma_X^2}\right].$$

$$(3.3.19)$$

Introduction of (3.3.8) yields an equation for $a_{J^+}(m)$ and thus a relationship analogous to (3.3.14)

$$\Delta_I^m \phi = -\Delta_m^{wt}\phi_{J^+}^0 + \frac{RT}{F} \ln$$

$$\times \left[a_{J^+}(1) + \frac{(\beta_{KX^+}c_X/\gamma_{KX^+}\gamma_X) + (\beta_{KX_2^+}c_X^2/\gamma_{KX_2^+}\gamma_X^2)}{(\beta_{JX^+}c_X/\gamma_{JX^+}\gamma_X)(\beta_{JX_2^+}c_X^2/\gamma_{JX_2^+}\gamma_X^2)} \right] K_{exch} a_{K^+}(1)$$

$$+ \frac{RT}{F} \ln \left[\frac{(\beta_{JX^+}c_X/\gamma_{JX^+}\gamma_X) + (\beta_{JX_2^+}c_X^2/\gamma_{JX_2^+}\gamma_X^2)}{c_0} \right]. \quad (3.3.20)$$

If the diffusion potential in the membrane is neglected, this equation yields relationships for the membrane potential, for the ISE potential and for the selectivity coefficient. It is apparent that formation of complexes with various numbers of ions in the membrane does not affect the dependence of the ISE potential on the activities of the determinand and interferent according to the Nikolsky equation.

In conclusion, consider a similar system where the determinand is a divalent and the interferent a monovalent cation, J^{2+} and K^+. For the complex concentration, as a result of the electroneutrality condition (activity coefficients are set equal to one),

$$c_0 = a_{J^{2+}}(m)(2\beta_{JX^{2+}}c_X + 2\beta_{JX_2^{2+}}c_X^2)$$
$$+ a_{K^+}(m)(\beta_{KX^+}c_X + \beta_{KX_2^+}c_X^2). \quad (3.3.21)$$

Activity $a_{K^+}(m)$ is eliminated using (3.2.23). It can be seen that the solution is similar to that given in (3.2.22) to (3.2.27). The resultant equation for the membrane/analyte potential yields an expression for the selectivity coefficient,

$$k_{J^{2+},K^+}^{pot} = \frac{(\beta_{KX^+}c_X + \beta_{KX_2^+}c_X^2)K_{exch}^{1/2}}{[2c_0(\beta_{JX^{2+}}c_X + \beta_{JX^{2+}}c_X^2)]^{1/2}} \quad (3.3.22)$$

related to the Nikolsky equation in the form

$$E_{ISE} = const + (RT/F) \ln (a_{J^{2+}}^{1/2} + k_{J^{2+},K^+}^{pot} a_{K^+}). \quad (3.3.23)$$

A survey of selectivity coefficients for ISEs with liquid membranes is found in the Appendix.

3.4 Membranes of solid and glassy materials

This group of membranes is also frequently called membranes with immobile ion-exchange sites. First systems in which *a diffusion potential is formed in the membrane* will be considered [10, 12, 18, 19, 25, 41, 54, 70]. It will be assumed that all sites are equivalent and that each has a single negative charge, while the membrane is permeable only for cations. It will be assumed also that two types of univalent cations are present in solution 1, J^+ and K^+, while solution 2 contains only ion J^+, and that both ions can enter the membrane and act as counter-ions for the anionic sites.

Exchange reaction (3.2.13) takes place between the ions in solution and those in the membrane. The procedure is the same as for (3.2.30) to (3.2.33), so that (assuming that the activity coefficients of the ions are equal to one in the membrane) an equation is obtained for the membrane potential:

$$\Delta\phi_M = \frac{RT}{F} \ln \frac{a_{J^+}(1) + (U_{K^+}/U_{J^+})K_{exch}a_{K^+}(1)}{a_{J^+}(2)}. \tag{3.4.1}$$

Eisenman [25] also gave a solution for the special case of type n membranes, where the ion activity is related to its concentration by the expression

$$\frac{d \ln a_{J^+}(m)}{d \ln c_{J^+}(m)} = \frac{d \ln a_{K^+}(m)}{d \ln c_{K^+}(m)} = n, \tag{3.4.2}$$

where n is independent of the distance from the membrane surface, so that

$$a_{J^+}(m) \sim c_{J^+}(m)^n. \tag{3.4.3}$$

The resultant expression for the membrane potential is then

$$\Delta\phi_M = \frac{nRT}{F} \ln \frac{a_{J^+}(1)^{1/n} + K_{exch}^{1/n}a_{K^+}(1)^{1/n}}{a_{J^+}(2)^{1/n}}. \tag{3.4.4}$$

A basic inadequacy of these solutions for membranes with immobile ion-exchange sites lies in the fact that a stationary state is assumed in the distribution of the diffusion potential across the membrane. This, of course, cannot be attained because of the considerable membrane thickness and slow transport through the solid phase. A further inadequacy is the assumption of constant mobility of the ions in the membrane. This assumption can probably be fulfilled inside the membrane, where a constant electric field resulting from the regularly distributed ion charges can be expected. However, in the vicinity of the membrane/electrolyte solution boundary the solvent effect will become important and will to a certain degree lead to hydration of the boundary surface layer. It will probably correspond to the real situation if the region in which a diffusion potential gradient is formed is identical with this hydrated layer.

Solid membranes without an internal diffusion potential [3, 8, 9, 54, 56, 62, 63] contain a nonporous layer of a poorly soluble salt in contact with a solution that contains either a cation or anion that is also a membrane component. A good example is silver halide membranes in solutions of either halide or silver ions.

Anionic response. The membrane potential depends on the activity of anions in the solution and Ag^+ acts as a charge carrier in the membrane. The behaviour of these membranes, at least in simple systems, is identical with the behaviour of the corresponding electrodes of the second kind, as will be shown below [43].

Consider an electrochemical membrane formed of poorly soluble salt JA, separating two solutions with different concentrations of anion A^-, c_2 and c_1. The different activities of these ions lead to the formation of a membrane potential, while no diffusion potential is formed inside the membrane. Under these conditions, for equilibrium between the membrane and solution 1, we have

$$\bar{\mu}_{J^+}(1) = \bar{\mu}_{J^+}(m)$$
$$\bar{\mu}_{A^-}(1) = \bar{\mu}_{A^-}(m),$$
(3.4.5)

that is,

$$\mu_{J^+}^0(wt) + RT \ln a_{J^+}(1) + F\phi(1) = \mu_{J^+}^0(m) + F\phi(m)$$
$$\mu_{A^-}^0(wt) + RT \ln a_{A^-}(1) - F\phi(1) = \mu_{A^-}^0(m) - F\phi(m),$$
(3.4.6)

where wt again designates the aqueous solution and m the membrane. Addition of (3.4.6) and combination with

$$\mu_{J^+}^0(m) + \mu_{A^-}^0(m) = \mu_{JA}^0$$
(3.4.7)

leads to

$$\mu_{JA}^0 - \mu_{J^+}^0(wt) - \mu_{A^-}^0(wt) = RT \ln [a_{J^+}(1)a_{A^-}(1)]$$
$$= RT \ln P_{JA},$$
(3.4.8)

where P_{JA} is the solubility product of salt JA.

As similar relationships hold for the second membrane/solution boundary, the equation for the membrane potential can be obtained from (3.4.6):

$$\Delta\phi_M = \phi(2) - \phi(1) = -(RT/F) \ln (a_{A^-}(1)/a_{A^-}(2)).$$
(3.4.9)

The membrane potential thus has the same value as the electric potential difference between two solutions 1 and 2 in a concentration cell with two electrodes of the second kind

$$\begin{array}{c|c|c|c|c|c|c}
\text{reference} & A^-(c_1) & JA & \text{metal} & JA & A^-(c_2) & \text{reference} \\
\text{electrode} & & & J & & & \text{electrode}.
\end{array}$$
(3.4.10)

If the activity of ion A^- is so low that it approaches the activity of this ion resulting from membrane dissolution, the same procedure used to derive (3.2.8) yields the membrane potential equation in the detection limit region,

$$\Delta\phi_M = -\frac{RT}{F} \ln \frac{a_{A^-}^0(1) + (a_{A^-}^0(1)^2 + 4P_{JA})^{1/2}}{2a_{A^-}(2)}.$$
(3.4.11)

Cationic response. The equation for the membrane potential for higher concentrations of cation J^+ is obtained in a similar manner:

$$\Delta\phi_M = \frac{RT}{F} \ln \frac{a_{J^+}(1)}{a_{J^+}(2)} \tag{3.4.12}$$

and, for low concentrations,

$$\Delta\phi_M = \frac{RT}{F} \ln \frac{a_{J^+}^0(1) + (a_{J^+}^0(1)^2 + 4P_{JA})^{1/2}}{2a_{J^+}(2)} \tag{3.4.13}$$

Equilibrium with two kinds of anion forming an insoluble salt. Consider membrane JA, where solution 2 contains only A^-, while solution 1 contains ions A^- as well as ions B^-, which also form an insoluble salt with J^+. If the activity of B^- is

$$a_{B^-} < (P_{JB}/P_{JA})a_{A^-}, \tag{3.4.14}$$

then the presence of B^- does not affect the membrane potential. If, on the other hand,

$$a_{B^-} > (P_{JB}/P_{JA})a_{A^-}, \tag{3.4.15}$$

then JA in the surface layer of the membrane is converted into JB, so that the new system can be depicted by the scheme

$$a_A\text{-}(1), a_B\text{-}(1) \left| \begin{array}{c} \text{JB} \\ \alpha \end{array} \right| \left| \begin{array}{c} \text{JA} \\ \beta \end{array} \right| a_A\text{-}(2). \tag{3.4.16}$$

When system (3.4.16) is at equilibrium,

$$\bar{\mu}_{J^+}(1) = \bar{\mu}_{J^+}(\alpha) = \bar{\mu}_{J^+}(\beta) = \bar{\mu}_{J^+}(2)$$
$$\bar{\mu}_{J^+}(1) + \bar{\mu}_{B^-}(1) = \mu_{JB}^0 \tag{3.4.17}$$
$$\bar{\mu}_{J^+}(2) + \bar{\mu}_{A^-}(2) = \mu_{JA}^0.$$

Finally, for the membrane potential we obtain

$$\Delta\phi_M = (1/F)[(\mu_{JB}^0 - \mu_{B^-}(1) - \mu_{B^-}(\text{wt})) - (\mu_{JA}^0 - \mu_{A^-}(1) - \mu_{A^-}(\text{wt}))]$$
$$= -\frac{RT}{F} \ln\left(\frac{P_{JB} \ a_{B^-}(1)}{P_{JA} \ a_{A^-}(2)}\right). \tag{3.4.18}$$

If it is assumed that adsorption of one of the insoluble salts on the surface of another does not occur and that no mixed salts are formed, the dependence of the membrane potential on the activity of one of the ions in the presence of the other will differ markedly from that depicted in fig. 3.2. Here, the resultant dependences should be linear with an intercept at point $a_{B^-} = (P_{JB}/P_{JA})a_{A^-}$ (see fig. 3.5) and the deviation from the Nernstian dependence should not be gradual as in fig. 3.2.

However, the dependence found experimentally shows this transition is usually gradual when the analyte contains two different types of halide ions.

Buck [8, 9] attempted to develop a theory describing this system and rightly pointed out that the transition from a dependence on A^- to a function depending on activity B^- is a result of formation of successive layers of JB until a structure with unit activity is formed. Nonetheless, it is not clear whether such a transition has equilibrium character.

For practical purposes it is appropriate to characterize the selectivities of membranes of similar insoluble salts (primarily silver halides) in terms of selectivity coefficients. These are, of course, only of empirical character and are found using (3.1.7). An approximation for the values of these constants is

$$k_{A^-, B^-}^{pot} \approx P_{JB}/P_{JA}. \qquad (3.4.19)$$

The effect on the ISE potential of diffusion of the product of the exchange

Fig. 3.5. Dependence of $\Delta\phi_M$ on the activity of I^-, at various activities of Cl^-, when the ions form insoluble Ag salts and there is no diffusion potential across the membrane. The solid line represents the theoretical prediction and the dotted line the commonly observed experimental dependence.

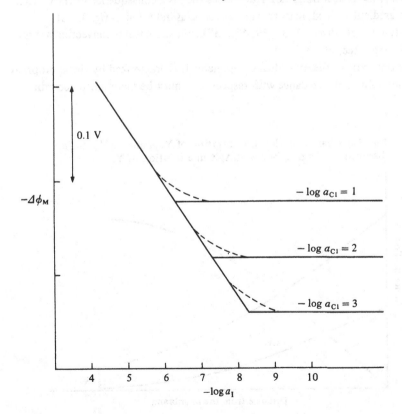

reaction in this kind of ISE is discussed theoretically by Hulanicki and Lewenstam [29, 30, 31].

Dissolution of membranes with complex formation [1, 7, 27, 38, 39, 44, 45, 49]. Consider a system in which a silver halide membrane AgX separates solution 1, containing halide ions at concentration c'_X and ions forming a complex with silver ion in concentration Y, from solution 2, containing halide ions at concentration c''_X. The exchange reaction

$$AgX + 2Y \rightleftharpoons AgY_2 + X \qquad (3.4.20)$$

occurs between ions Y and the silver halide, with an equilibrium constant defined by the equation (for simplicity the activity coefficients have been set equal to one)

$$K = c_{AgY_2} c_X / c_Y^2 = \beta_2(AgY_2) P_{AgX}, \qquad (3.4.21)$$

where $\beta_2(AgY_2) = c_{AgY_2} / (c_{Ag} c_Y^2)$.

During dissolution, the concentrations of X and AgY_2 increase at the phase boundary, the concentration of Y decreases and, as a consequence of the concentration gradient formed, mass transport occurs, as indicated in fig. 3.6. If the electrolyte is not stirred [1, 38, 39, 40], diffusion and natural convection are the only driving forces of transport.

The transport of the individual components is characterized by the appropriate mass flux values. Mass balance with respect to Y must be maintained according

Fig. 3.6. Changes in the concentration of Y, X and AgY_2^- during dissolution of a membrane of AgX in a solution of Y.

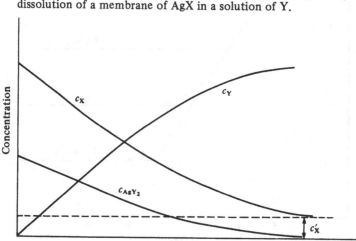

to the equation

$$J_Y + 2J_{AgY_2} = 0,$$ (3.4.22)

where the Js are the mass fluxes.

Charge balance must, of course, also be maintained (as X, Y and AgY_2 have a single negative charge),

$$J_{AgY_2} = J_X.$$ (3.4.23)

The mass fluxes are expressed by the products of the convection–diffusion constants, κ_X, κ_Y and κ_{AgY_2}, and the differences between the concentrations of the substances in the bulk of the solution far from the membrane (c'_X, c_Y and c_{AgY_2}, of which the last equals zero) and the concentrations immediately next to the membrane $c_X(0)$, $c_Y(0)$ and $c_{AgY_2}(0)$.

$$\kappa_Y[c_Y - c_Y(0)] - 2\kappa_{AgY_2} c'_{AgY_2}(0) = 0$$
$$\kappa_{AgY_2} c_{AgY_2}(0) = \kappa_X[c_X(0) - c'_X].$$ (3.4.24)

It is then necessary to assign a numerical value to $c_X(0)$, which can then be substituted into the equation for the membrane potential (3.4.9)

$$\Delta\phi_M = (RT/F) \ln [c''_X/c_X(0)].$$ (3.4.25)

Using (3.4.24), $c_{AgY_2}(0)$ and $c_Y(0)$ are calculated and substituted in the modified (3.4.21)

$$K = c_{AgY_2}(0)c_X(0)/c_Y(0)^2.$$ (3.4.26)

In this way, the following equation involving $c_X(0)$ is obtained:

$$K = \frac{[(\kappa_X/\kappa_{AgY_2})c_X(0) - (\kappa_X/\kappa_{AgY_2})c_X]c_X(0)}{[-(2\kappa_X/\kappa_Y)c_X(0) - (2\kappa_X/\kappa_Y)c_X + c_Y]^2}.$$ (3.4.27)

If the constant K has a value much larger than unity, practically all ions of the complexing agent approaching the membrane react with AgX, liberating X in an amount corresponding to one half of their amount. If, making this assumption, the individual terms in the sums in the numerator and the denominator of (3.4.27) are compared, it can be seen that the denominator in this equation will always be a small quantity. Thus we can write

$$(-(2\kappa_X/\kappa_Y)c_X(0) + (2\kappa_X/\kappa_Y)c_X + c_Y)^2$$
$$= (1/K) [(\kappa_X/\kappa_{AgY_2})c_X(0) - (\kappa_X/\kappa_{AgY_2})c_X]c_X(0)] \approx 0.$$ (3.4.28)

It then follows that

$$c_X(0) = c'_X + (\kappa_Y/2\kappa_X)c_Y = c'_X + \tfrac{1}{2}(D_Y/D_X)^\beta c_Y,$$ (3.4.29)

where D_X and D_Y are the diffusion coefficients of X and Y respectively, and β is a constant with a value of about 0.6 to 0.75. The latter value arises from the theory of transport under natural convection, with a vertically placed phase boundary [32].

The approximation (3.4.28) is fulfilled for dissolution of all silver halides in cyanide ion solutions. Even in the case of the least soluble halide, AgI, the solubility product equals 10^{-16}, while $\beta_{Ag(CN)_2}$ varies between $10^{18.42}$ and $10^{19.85}$ (see [5]).

After substitution from (3.4.29) into (3.4.25), the final result

$$\Delta\phi_M = \frac{RT}{F} \ln \frac{c_X''}{c_X' + \frac{1}{2}(D_Y/D_X)^\beta c_Y} \qquad (3.4.30)$$

is obtained.

It is, in this case, possible to speak in a formal sense about a selectivity constant having the value

$$k_{X,Y}^{pot} = \frac{1}{2}(D_Y/D_X)^\beta \approx \frac{1}{2}. \qquad (3.4.31)$$

It should be noted that the concentration of free ions of the complexing agent appears in (3.4.30). Since in this case we have cyanide ions, the method cannot be used at pH values below 9, when the complexing agent is largely undissociated, or in cases when its ions are bound in another complex. Then the value c_Y represents the total concentration of the complexing agent not bound in the complex ($c_Y = [Y^-] + [HY]$) [28, 44, 45, 48].

The constant K of (3.4.26) and (3.4.27) must then be replaced by the value

$$K' = KK_{HY}^2/(K_{HY} + [H^+])^2. \qquad (3.4.32)$$

In the special case when $c_X = 0$, (3.4.28) takes the form

$$-\frac{2\kappa_X}{\kappa_Y} c_X(0) + c_Y)^2 = \frac{\kappa_X c_X(0)^2 (K_{HY} + [H^+])^2}{\kappa_{AgY_2} K K_{HY}^2} \approx 0. \qquad (3.4.33)$$

Hence

$$c_X(0) = c_Y(\kappa_{AgY_2}(\kappa_X)^{1/2}$$

$$\times \frac{K^{1/2}K_{HY}/(K_{HY} + [H^+])}{1 + 2(\kappa_X\kappa_{AgY_2}/\kappa_Y^2)^{1/2}K^{1/2}K_{HY}/(K_{HY} + [H^+])^2}. \qquad (3.4.34)$$

When $K_{HY} \ll [H^+]$ and $K^{1/2}K_{HY}/[H^+] \ll 1$, from (3.4.25), the membrane potential dependence on pH is

$$\Delta\phi_M = -(RT/F) \ln c_Y[\kappa_{AgY_2}(\kappa_X)^{1/2}K^{1/2}K_{HY}/([H^+]c_X'')]$$

$$\approx -(RT/F) \ln [c_Y(D_{AgY_2}/D_X)^{\beta/2}K^{1/2}K_{HY}/([H^+]c_X'')]. \qquad (3.4.35)$$

References for Chapter 3

1 T. Aomi, *Denki Kagaku* **46**, 617 (1978).

2 S. Bäck and J. Sandblom, *Anal. Chem.* **45**, 1680 (1973).

3 F. G. K. Baucke, *Elektrochim. Acta* **17**, 851 (1972).
4 G. Baum, *J. Phys. Chem.* **76**, 1872 (1972).
5 J. Bjerrum, G. Schwarzenbach and L. G. Sillén, *Stability Constants of Metal-Ion Complexes, Part II: Inorganic Ligands*, The Chemical Society, London (1958), p. 35 and 119.
6 J. H. Boles and R. P. Buck, *Anal. Chem.* **45**, 2057 (1973).
7 G. P. Bound, B. Fleet, H. von Storp and D. H. Evans, *Anal. Chem.* **45**, 788 (1973).
8 R. P. Buck, *Anal. Chem.* **40**, 1432 (1968).
9 R. P. Buck, *Anal. Chem.* **40**, 1439 (1968).
10 R. P. Buck, *Anal. Chem.* **45**, 654 (1973).
11 R. P. Buck, *Hung. Sci. Instr.* **49**, 7 (1980).
12 R. P. Buck, *Theory and Principles of Membrane Electrodes, in Ion-Selective Electrodes in Analytical Chemistry* (ed. H. Freiser), Vol. I, Plenum Press, New York (1978), p. 1.
13 R. P. Buck and F. S. Stover, *Anal. Chim. Acta* **101**, 231 (1978).
14 R. P. Buck, F. S. Stover and D. E. Mathis, *J. Electroanal. Chem.* **82**, 345 (1977).
15 K. Cammann, *A Mixed Potential Ion-Selective Electrode Theory* (eds. E. Pungor and I. Buzás) International Conference held in Budapest 1977, Akadémiai Kiadó, Budapest and Elsevier, Amsterdam (1978).
16 K. Cammann, *Anal. Chem.* **50**, 936 (1978).
17 S. Ciani, G. Eisenman and G. Szabo, *J. Membrane Biol.* **1**, 1 (1969).
18 F. Conti and G. Eisenman, *Biophys. J.* **5**, 247 (1965).
19 F. Conti and G. Eisenman, *Biophys. J.* **5**, 511 (1965).
20 A. K. Covington and P. Davison, Liquid ion exchanger types, in *Ion-Selective Electrode Methodology*, Vol. I, CRC Press, Boca Raton (1979), p. 85.
21 P. R. Danesi, F. Salvemini, G. Scibona and B. Scuppa, *J. Phys. Chem.* **75**, 554 (1971).
22 E. Dubini-Paglia, R. Galli and T. Mussini, *Experientia Suppl.* **18**, 259 (1971).
23 G. Eisenman, Bioelectrodes, in *Modern Techniques of Physiological Sciences* (ed. J. F. Gross), Academic Press, New York (1974), p. 245.
24 G. Eisenman, *Anal. Chem.* **40**, 310 (1968).
25 G. Eisenman, Theory of membrane electrode potentials: an examination of the parameters determining the selectivity of solid and liquid ion exchangers and of neutral sequestering molecules, Chapter 1 of *Ion-Selective Electrodes* (ed. R. A. Durst), National Bureau of Standards, Washington (1969).
26 G. Eisenman, S. Ciani and G. Szabo, *J. Membrane Biol.* **1**, 294 (1969).
27 D. H. Evans, *Anal. Chem.* **44**, 875 (1972).
28 M. Gratzl, F. Rakiás, G. Horvai, K. Tóth and E. Pungor, *Anal. Chim. Acta* **102**, 85 (1978).
29 A. Hulanicki and A. Lewenstam, Interpretation of the Response Mechanism of Solid-State Ion-Selective Electrodes Regarding Diffusion Processes, in *Ion-Selective Electrodes* (ed. E. Pungor and I. Buzás), International Conference held in Budapest 1977, Akadémiai Kiadó, Budapest and Elsevier, Amsterdam (1978).
30 A. Hulanicki and A. Lewenstam, *Talanta* **26**, 661 (1977).
31 A. Hulanicki and A. Lewenstam, *Talanta* **24**, 171 (1976).
32 N. Ibl, Application of mass transfer theory. The formation of powdered metal deposits, in *Advances in Electrochemistry and Electrochemical Engineering* (eds. P. Delahay and C. W. Tobias), Vol. 2, J. Wiley and Sons, New York (1962), p. 58.
33 N. Ishibashi, H. Kohara and N. Uemura, *Bunseki Kagaku* **21**, 1072 (1972).
34 H. J. James, G. P. Carmack and H. Freiser, *Anal. Chem.* **44**, 853 (1972).

35 H. J. James, G. P. Carmack and H. Freiser, *Anal. Chem.* **44**, 856 (1972).
36 A. Jyo, K. Fukamachi, W. Koga and N. Ishibashi, *Bull. Chem. Soc. Japan* **50**, 670 (1977).
37 A. Jyo, M. Tarikai and N. Ishibashi, *Bull. Chem. Soc. Japan* **47**, 2862 (1974).
38 W. Jaenicke, Z. *Elektrochemie* **55**, 648 (1951).
39 W. Jaenicke, Z. *Elektrochemie* **57**, 843 (1953).
40 W. Jaenicke and M. Haase, Z. *Elektrochemie* **63**, 521 (1959).
41 G. Karrenman and A. G. Eisenman, *Bull. Math. Biophys.* **24**, 413 (1962).
42 O. Kedem, M. Perry and R. Bloch, *Charged React. Polym.* **4**, (Charged Gels Membr. Part 2) 125 (1976).
43 I. M. Kolthoff and H. L. Sanders, *J. Am. Chem. Soc.* **59**, 416 (1937).
44 J. Koryta, *Anal. Chim. Acta* **61**, 329 (1972).
44*a* J. Koryta, *Anal. Chim. Acta* **111**, 1 (1977).
45 J. Koryta, *Iontově-selektivní membránové elektrody*, Academia, Praha (1971).
46 J. Koryta, *Hung. Sci. Instr.* **49**, 25 (1980).
46*a* J. Koryta, J. Dvořák and V. Boháčková, *Electrochemistry*, Chapman & Hall, London (1973).
47 A. A. Lev, V. V. Malev and V. V. Osipov, *Membranes – A Series of Advances* (ed. G. Eisenman), Vol. 2, M. Dekker, New York (1973), p. 479.
48 M. Mascini, *Anal. Chem.* **45**, 614 (1973).
49 M. Mascini and A. Napoli, *Anal. Chem.* **46**, 447 (1974).
50 D. E. Mathis, R. M. Freeman, S. T. Clark and R. P. Buck, *J. Membrane Sci.* **5**, 103 (1979).
51 C. McCallum and R. Paterson, *J. Chem. Soc. Faraday Trans.* 1, **72**, 323 (1976).
52 K. H. Meyer and J. F. Sievers, *Helv. Chim. Acta* **19**, 649, 665 (1936).
53 K. H. Meyer and J. F. Sievers, *Trans. Faraday Soc.* **33**, 1073 (1937).
54 W. E. Morf, *The Principles of Ion-Selective Electrodes and of Membrane Transport*, Akadémiai Kiadó, Budapest (1981).
55 W. E. Morf, D. Ammann and W. Simon, *Chimia* **28**, 65 (1974).
56 W. E. Morf, G. Kahr and W. Simon, *Anal. Chem.* **40**, 1538 (1974).
57 W. E. Morf, G. Kahr and W. Simon, *Anal. Lett.* **7**, 9 (1974).
58 B. P. Nikolsky, *Zh. Fiz. Khim.* **10**, 495 (1937).
59 B. P. Nikolsky and M. M. Shults, *Zh. Fiz. Khim.* **36**, 1327 (1962).
60 M. Perry, E. Löbel and R. Bloch, *J. Membrane Sci.* **1**, 223 (1976).
61 B. C. Pressman, *Proc. Nat. Acad. Sci.* **53**, 1076 (1965).
62 E. Pungor, *Anal. Chem.* **39**, 28A (1967).
63 E. Pungor and K. Tóth, *Analyst* **95**, 625 (1970).
64 O. Ryba and J. Petránek, *J. Electroanal. Chem.* **67**, 321 (1976).
65 J. P. Sandblom, *J. Phys. Chem.* **73**, 249 (1969).
66 J. P. Sandblom, G. Eisenman and J. L. Walker Jr., *J. Phys. Chem.* **71**, 3862 (1967).
67 J. P. Sandblom, G. Eisenman and J. L. Walker Jr., *J. Phys. Chem.* **71**, 3871 (1967).
68 J. P. Sandblom and F. Orme, Liquid membranes as electrodes and biological models, in *Membranes – A Series of Advances* (ed. G. Eisenman), Vol. I, M. Dekker, New York 1972, p. 125.
69 R. Scholer and W. Simon, *Helv. Chim. Acta* **55**, 1081 (1972).
70 M. M. Shults, *Dokl. AN SSSR* **194**, 337 (1970).
71 M. M. Shults and O. K. Stefanova, *Vestnik Leningrad. Univ.* **22** (1971).
72 Z. Štefanac and W. Simon, *Chimia* **20**, 436 (1966).
73 F. S. Stover and R. P. Buck, *Biophys. J.* **16**, 753 (1976).
74 F. S. Stover and R. P. Buck, *J. Electroanal. Chem.* **94**, 59 (1978).

75 G. Szabo, G. Eisenman and S. Ciani, Ion distribution equilibria in bulk phases and the ion transport properties in lipid bilayer membranes produced by neutral macrocyclic antibiotics, in *Proc. Coral Gables Conf. on the Physical Principles of Biological Membranes*, 1968, Gordon and Breach, New York (1969).

76 T. Teorell, *Trans. Faraday Soc.* **33**, 1053 (1937).

77 T. Teorell, *Z. Elektrochemie* **55**, 460 (1951).

78 A. P. Thomas, A. Viviani-Nauer, S. Arvanitis, W. E. Morf and W. Simon, *Anal. Chem.* **49**, 1567 (1977).

79 J. L. Walker, G. Eisenman and J. P. Sandblom, *J. Phys. Chem.* **72**, 978 (1968).

4

PRINCIPAL PROPERTIES OF
ION-SELECTIVE ELECTRODES

As stated on p. 28, an analytically ideal sensor would determine the determinand both specifically and quantitatively. In potentiometry, this would require an electrode sensitive to one single substance among all the components of the system.

Cationic metal electrodes definitely do not have this property. They attain (though not always) a potential that is a function of the activity of the corresponding ion according to the Nernst equation; however, in the presence of the ions of nobler metals these electrodes become covered with a layer of the deposited nobler metal and thus have quite different properties. Again, in the presence of components of oxidation–reduction systems in the test solution, mixed potentials develop at these electrodes (see for example [86a], and p. 000). The dependence of this potential on the solution composition is complex and thus such electrodes are unsuitable for analytical purposes. Some cation electrodes do not obey the Nernst equation with respect to the corresponding ions and their potentials are simultaneously affected by the formation of an oxide film and by reactions with various solution components (hydrogen ions, oxygen, etc.). Therefore, cation electrodes are not suitable sensors for specific estimation of the determinand concentration. In optimal cases, redox electrodes sometimes yield thermodynamic potentials corresponding to components in the solution, but mixed potentials are also often encountered.

Anion electrodes of the second kind have better properties in this respect. They respond to the corresponding anions over wide activity ranges and the only major interference with their function comes from anions forming salts less soluble than the salt covering the electrode. However, strong oxidants also interfere.

As has been shown over the last fifteen years, the membrane systems of ion-selective electrodes (ISEs) approach specificity under certain conditions. In this book, an ion-selective electrode is defined as a sensor with a membrane whose potential indicates the determinand activity. The membranes of ion-selective electrodes consist of liquid electrolyte solutions or of solid or glassy electrolytes that usually have negligible electron conductivity under the given conditions. Attempts have been made to include all potentiometric sensors among ion-selective electrodes. We feel that this broadening of the concept of ion-selective electrodes is unsuitable, because the above definition specifies a clearly limited set of potentiometric sensors and, moreover, is generally used.

Ion-selective electrodes have certain undoubted advantages: (*a*) they do not affect the test solution; (*b*) they are portable; (*c*) they are suitable both for direct determinations and as sensors in titrations; (*d*) they are not expensive. Here it should be emphasized again that there is a difference between the concepts 'specific' and 'selective', because these are often used imprecisely. Specificity is characterized by the fact that the membrane potential is affected by a single kind of ion in solution (see the previous chapter). On the other hand, the potential of a selective membrane is affected by more than one kind of ion, but a certain kind is preferred in the response.

It seems surprising that systems of water-permeable polymeric ion-exchange membranes with ion-exchanging groups built directly into the polymeric structure have not found wider use here. Two groups of systems gave good results in the development of ion-selective electrodes: (*a*) electrodes with fixed ion-exchange sites in a solid or glassy membrane structure; (*b*) electrodes with liquid membranes. In the former system, a time-independent cross-linked structure of a certain kind of ion is formed in the membrane. Such a membrane is either homogeneous (single crystal, polycrystalline substance or glass), or heterogeneous (a crystalline substance is built into a skeleton of a suitable polymer). In the latter system, a water-immiscible liquid forms the membrane, in which a salt of the determinand with a strongly hydrophobic ion-exchanger ion or a complex of the determinand with a hydrophobic complexing agent (ion-carrier or ionophore) is dissolved. This solution is either used as a polymeric film plasticizer, or a porous diaphragm is soaked with it.

This chapter deals with ISE construction, their characteristic properties such as selectivity coefficient, response time, temperature coefficient and drift, as well as electrode calibration and composite sensors containing ISEs.

4.1 Construction of ion-selective electrodes

Here the construction of macroelectrodes and the possibility of replacing the internal ISE solution by a metallic contact will be discussed. The principal requirement is that the membrane completely separates the test solution from the electrode interior, because otherwise irregular deviations from the calibration dependence occur.

Contemporary potentiometers contain circuits with high internal resistances. The resistance of the electrode membranes thus creates no serious problem, but the leads to the electrodes should always be shielded and the high-resistance systems should be placed in a Faraday cage; otherwise the measurement is subject to noise.

Macroelectrodes with solid membranes contain homogeneous [142] or heterogeneous [25] membranes. The construction of an ISE of this type with an internal reference electrode is shown in fig. 4.1. For good functioning of an ISE it is necessary that the membrane be completely sealed in the electrode body, with no cracks leading to short-circuiting between the external and internal solutions. Cements based on Teflon, PVC or epoxy resin are used [170].

Homogeneous membranes [142] are based on single crystals [45], solidified salt melts [86], pressed pellets of salt powders [142], or ceramic systems obtained by sintering and pressing at high temperatures [61, 63, 64, 65]. An electrode with a rotating membrane was proposed for studying this type of electrode [60]. Membranes obtained by sintering of mixtures of conductive materials (for example AgCl and Ag_2S) will also be included among homogeneous membranes, although they are internally heterogeneous. Pellets are obtained by pressing at

Fig. 4.1. Illustrative diagram of a solid-membrane ISE.

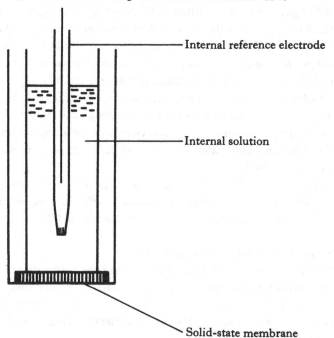

Internal reference electrode

Internal solution

Solid-state membrane

pressures from 7×10^7 to 2×10^8 Pa [169]. A noteworthy homogeneous solid
membrane modification is an electrode made from a split crystal [187]. This
membrane is prepared by splitting a thin disc-shaped single-crystal membrane
into two semicircular identical membranes that are pretreated identically. Two
sensors with identical properties are thus obtained and are suitable, for example,
for differential titrations. Electrodes with an internal metallic contact (all-
solid-state) are used wherever possible. The preparation of electrodes of this
type from fused membrane materials is described in [159]. Similar electrodes
consisting of a thin layer of the membrane material deposited on a suitable
ionic conductor have also been proposed [171, 172]; Ag_2S, Ag_3SBr, Ag_3SI or
$Ag_{19}I_{15}P_2O_7$ are used as the conductive supports. The adverse effect of
adsorption at the boundaries of the individual particles in the membrane was
also investigated using these electrodes [173].

The glass electrode also belongs among homogeneous membranes. For the
sake of completeness, its properties will be described in chapter 6, but for
construction details the reader is referred to the literature ([48]).

Heterogeneous membranes are surveyed in [25]. Pungor played a pioneering
role in the development of ISEs with his membranes of precipitates of sparingly
soluble salts built into an inert matrix (for a survey see [29, 122, 126, 127]).
Homogeneous membranes, as described in the previous paragraph, do have
definite advantages over heterogeneous membranes in their better reproducibility
(see [17a]), but the preparation of homogeneous membranes requires the use
of special techniques, as described in chapter 6, whereas precipitate membranes
can be prepared in ordinary laboratories. The following procedure is used by
Pungor and coworkers [122, 124] (see also [29, 136]) for the preparation
of electrodes with a silicone rubber matrix. A mixture of the precipitate and
polysiloxane is homogenized, a cross-linking agent (a silane derivative) and
a catalyst are added so that the mixture contains about 50% precipitate. The
required membrane shape is obtained by calandering. The quality of the membrane depends on the degree of cross-linking, because it determines the distribution of the precipitate particles in the membrane. Buchanan and Seago [19]
recommend mixing silicone rubber with pulverized silver halides and pressing
the mixture between a polyethylene plate and a PVC foil. Other procedures
involve precipitation of silver halides in a thermoplastic polymer [102, 174] or
in a polyethylene matrix [91, 103]. In the latter case, the membranes must be
pressed at 100 to 130 °C and at a pressure of 10^7 to 3×10^7 Pa. Dentacryl is
also a suitable material for the membrane matrix [175].

Hirata and Date [62] built an insoluble precipitate (Cu_2S) into a silicone
rubber or epoxide resin matrix and deposited the mixture on a metal wire, thus

obtaining a simple ISE. The electrode functioned only when the precipitate particles were in contact in the membrane.

The 'Selectrode' [54, 144, 146, 147, 148] also belongs among heterogeneous membrane systems. The electrode body consists of a pressed rod of graphite hydrophobized with Teflon, contained in a Teflon tube (see fig. 4.2). The electrode contact is a shielded stainless steel wire, screwed into the rod. By rubbing 2–4 mg of finely powdered (or colloidal) material into the graphite surface, a firm membrane is obtained. The loose material is then removed and the electrode is polished manually. When the membrane loses its activity, the upper layer is cut off with a scalpel and a fresh surface is obtained, which can again be covered with an active material. In spite of the simple preparation procedure, these electrodes function relatively well. For the problem of the contact of the membrane material with graphite see p. 64.

Liquid-membrane macroelectrodes come in two completely different versions. The older type, no longer extensively used, contains a porous

Fig. 4.2. Růžička's 'Selectrode': 1 – sensitive surface; 2 – stainless steel contact; 3 – screening; 4 – Teflon tubing; 5 – cylinder pressed from graphite hydrophobized by Teflon; 6 – electroactive material. (After J. Růzicka, C. G. Lamm and J. C. Tjell.)

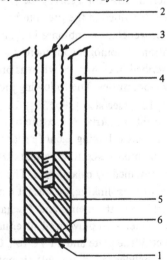

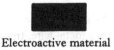
Electroactive material

Hydrophobized graphite

diaphragm soaked with an ion-exchanger solution closing the electrode body and thus separating the internal solution of the electrode from the test solution. The diaphragm is connected by channels with the ion-exchanger solution reservoir, thus providing for constant replenishment of the liquid membrane [133] For the diaphragm material, porous Teflon, sintered ceramic filters with siliconizec pores or glass frits have been used. The main drawback of this electrode is bleeding of the ion-exchanger solution into the test solution, which necessitates continuous replenishment by transport from the membrane edges where the ion-exchanger solution reservoir is placed. Another source of problems is the difficulty in completely filling the membrane pores with the ion-exchanger solution.

The above difficulties are removed in the new version of the liquid membrane, which employs a polymeric film with the ion-exchanger solution functioning as a plasticizer. Then it is much easier to prepare a membrane without leaks and using only a minute amount of the ion-exchanger solution. When the membrane ceases to function, it is simply replaced. For a survey of those electrodes see [109, 111, 112, 113, 180]; they are generally termed solvent–polymeric membranes [180] or polyvinyl chloride–matrix membranes [112].

The idea of using a polymeric membrane with an ion-selective plasticizer originated with Shatkay and coworkers [14, 72, 151] who, in a search for a suitable system for a calcium ion-selective electrode, used thenoyltrifluoro-acetone in tributylphosphate as an ion-exchanger solution in a PVC membrane. This system was unsuccessful, because of insufficient selectivity. Kedem and coworkers [82] patented the membrane preparation procedure using a mixture of a cellulose acetate solution in acetone and a solution of a suitable ionophore in dimethylsebacate. When this solution dries on a glass plate, a film containing the ionophore and the plasticizer at a very fine dispersion is obtained. Because the cationic complex of an ionophore (for example the potassium complex of valino-mycin) is positively charged, the polymer must have a weak negative charge.

The method of Moody, Thomas and coworkers [31, 33, 34, 49, 58, 108] has enjoyed general acceptance. The simple procedure involves pouring of a mixture of 0.4 g ion-exchanger solution and 0.17 g PVC in 6 cm^3 tetrahydrofuran onto a polished glass plate, on which a circular cylinder with a diameter of 30-35 mm and a height of 30 mm is placed to prevent the solution from spreading. Cyclo-hexanone is sometimes used instead of tetrahydrofuran. The mixture is allowed to evaporate at laboratory temperature, with a cylinder covered by an unweighted filter paper. The solution dries after about two days and the membrane obtained is about 0.2 mm thick. Discs of a suitable size (about 5 mm in diameter) are cut from this membrane, cemented to PVC tubes with a PVC cement in tetrahydro-furan and the PVC tubes are inserted over glass tubes, thus obtaining electrode bodies. These are filled with the internal solution and provided with an internal reference electrode. The dispersion of the ion-exchanger solution in the PVC film is shown in fig. 4.3.

The dropping ISE described by Skobets and coworkers [160, 161], where the ion-exchanger solution drops from a glass capillary, permits regular renewal of the electrode surface. A nitrate ISE with a renewable surface [162a] is shown in fig. 4.4.

A special type of ISE is used in the two-phase titration [17, 50, 51, 52], in which the titrant (for example picric acid) is added to the aqueous phase containing the test ion (for example the trioctylethylammonium ion), which is in contact with a nitrobenzene phase containing picrate ions and functioning as an ISE.

Plastic film membranes can also contain fixed ion-exchange groups. Jyo and coworkers [79] chloromethylated Amberlite XAD-2 (cross-linked styrene-divinylbenzene copolymer of the macroreticular type) and formed quaternary ammonium groups in the product by treatment with dimethyltetradecylamine. They converted the substance into the chloride, nitrate or perchlorate form and saturated it with nitrobenzene. The presence of hydrophobic ion-exchange sites

Fig. 4.3. Visible-light microscope photomicrograph showing pores (dark circles) in polyvinylchloride matrix membrane incorporating didecylphosphoric acid-dioctylphenylphosphonate shaken with a molar solution of calcium chloride. (From [49]. By permission of the Society of Analytical Chemistry, London.)

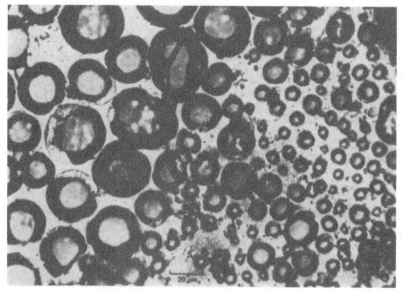

and of non-aqueous solvent renders the membrane selective in a way considerably different from the selectivity of hydrophilic ion-exchanger membranes. Keil and coworkers [83] phosphorylated the VAGH copolymer of vinylchloride and vinylacetate by using decyldihydrogenophosphate, converted the product into the calcium salt and mixed it with dioctylphenylphosphonate and PVC. The membrane obtained did not exhibit any pronounced advantages over the common type of Ca^{2+}ISE. Ebdon *et al.* [36, 37] grafted triallylphosphate to PVC or VAGH, converted the phosphorylated copolymer into the calcium salt after partial hydrolysis and made a polymeric film from it without a filling compound or with Orion 92-20-02 liquid ion-exchanger as the filler. Thus they obtained ISEs with quite good properties but not, however, better than those of common solvent–polymeric membranes.

Fig. 4.4. Nitrate ISE with renewable surface: 1 – outer body of ISE; 2 – inner jacket; 3 – contact to inner reference electrode; 4 – stopcock; 5 – ion-exchanger solution; 6 – reservoir for the ion-exchanger solution; 7 – orifice with ion-sensitive interface; 8 – glass frit; 9 – connecting PVC tube; 10 – inner solution of ISE; 11 – analyte. (After Šenkýř and Petr [149*a*].)

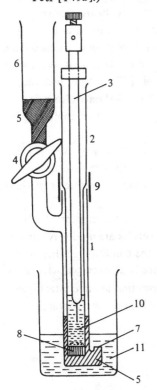

A Selectrode with a plastic film [41, 145] has also been constructed. The membrane does not directly touch the graphite rod, but is separated by a layer consisting of saturated KCl, mercury and calomel (with valinomycin K^+ ISE). An ISE with a plastic membrane directly attached to a graphite rod is described in [4].

Coated-wire electrodes consist of a plastic membrane deposited directly on a metal wire (originally of platinum, but later it was found that copper lead is sufficient) [25, 26, 46, 47, 73, 110, 182, 186]. For the preparation of such a very simple sensor [100] it is sufficient to remove the insulation from a coaxial copper cable and to free the copper core from the insulating layer for a length of about 2 cm. The wire is carefully cleaned, dried and immersed several times about 1 cm into a mixture of PVC and plasticizer solutions (similar to the preparation of solvent–polymeric membranes. It is allowed to dry for about one minute in a vertical position. The procedure is repeated several times until a bead with a diameter of about 2 mm is formed at the end of the wire. The electrode is allowed to dry overnight and the wire, except for the bead, is suitably insulated. The choice and amount of plasticizer are significant for the membrane properties, because an insufficient amount of a plasticizer does not lower the glass-transition temperature, T_g, of the polymer below laboratory temperature, T_v, a lowering which is required if the membrane is to function well. The electrode is then stored overnight in a 0.1 M solution of a determinand salt. When used frequently it can be stored in the same solution saturated with the ion-exchanger; otherwise it can be kept in the air [47]. A 'split membrane' coated-wire ISE [25a] and miniature coated-wire ISEs [43] have also been described. Coated-disk ISEs have been proposed [154, 156] for monitoring in clinical analysis.

The graphite paste electrode, introduced into voltammetry by Adams [1], has an analogue in an ISE where a mixture of graphite powder with an ion-exchanger solution functions as the electrode membrane, into which a platinum contact is immersed [16, 129, 149].

We now come to internal metal contacts in ISEs without an internal solution. As discussed above, systems without internal electrolytes are used very often, with both solid and liquid membranes. Obviously, the condition of thermodynamic equilibrium requires that common electrically-charged particles (ions or electrons) be present in electrically-charged phases that are in contact (see chapter 2). ISEs with a silver halide membrane to which a silver contact is attached are relatively simple. In the system

1	2	3	4
JX, H_2O	AgX	Ag	Cu

ion X^- is common to phases 1 and 2, ion Ag^+ to phases 2 and 3, and phases 3 and 4 (the leads to the potentiometer) have electrons in common. Therefore, the condition for thermodynamic equilibrium is satisfied. Somewhat different conditions are encountered when another conductor is in contact with phase 2, for example graphite. Here it is necessary to consider the composition of membrane 2 if the measurement is to yield unambiguous results (see chapter 6). Obviously it is unsuitable to measure the 'standard potential' of an ISE in the system [101]

Ag | AgX | JX,H_2O | AgX-membrane | Hg | Pt.

It is incorrect to vacuum-plate the internal side of the membrane with silver or gold in the fluoride ISE containing the lanthanum trifluoride membrane [177]. It is evidently better to use the system [42, 96, 117]

LaF_3 | AgF | Ag | Cu.

When LaF_3 is in direct contact with metals, such as Ag, Hg, Pb, Ga, Cd, $ZnHg_x$ or La, the contact attains the properties of an ideally polarized interface [96] with poorly defined E_{ISE} values.

As shown in the previous section, many liquid-membrane electrodes contain a metal or graphite in direct contact with the membrane. Buck [21] considers the liquid PVC membrane/metal contact interface as a kind of 'blocked' phase boundary, i.e. basically an ideally polarized boundary. However, at such a boundary a constant potential difference required for a defined ISE potential cannot exist and the potential difference is a function of the electrode charge, which can be changed by bringing charge from an external source or by undefined electrode reactions. The other, much more likely possibility is the formation of a mixed potential at this boundary [86a], in the system consisting of the metal, metal oxide, oxygen, ion-exchanger solution or some components of the polymeric system. This mixed potential is probably constant under certain conditions and thus the ISE potential does not significantly depend on the conditions at the membrane/internal contact interface. Nevertheless, the effect of oxygen on platinum, copper and silver internal contacts is distinctly manifested in the ISE potential. The best results have been obtained with internal contacts consisting of a silver wire coated with AgCl [57, 69, 155, 157].

4.2   Ion-selective microelectrodes

Ion-selective microelectrodes [18, 70, 71, 164] are chiefly used for measurement of ion activities in individual cells and in intracellular liquid. They were developed from micropipettes, which are miniature liquid bridges used for measurement of cell membrane potentials [94]. Micropipettes and ion-selective microelectrodes are made using commercial drawing devices. Ion-selective microelectrodes (ISM) are of three types: glass electrodes (for measurement intracellular pH and sodium ion activity), solid-membrane ISMs (for measuring the chloride ion activity) and liquid-membrane ISMs (especially for measuring the activities of potassium, chloride and calcium ions).

For the sake of completeness, glass microelectrodes [48, 59, 184] will first be mentioned. Two types of these electrodes are used, spear-shaped micro-electrodes [59] and recessed-tip microelectrodes [165] (see fig. 4.5). In the former case, the microelectrode is drawn from a capillary of an ion-exchanger glass and is insulated on the outside, except for the tip, by inserting the microelectrode into a micropipette made of an inactive glass. In the latter case, the outer micropipette extends over the microelectrode tip. The two capillaries are sealed together and the ISM is in contact with the liquid between the two capillaries.

Fig. 4.5. Two types of ion-sensitive glass microelectrodes: (*a*) Hinke's spear type; (*b*) recessed-tip type. (After J. L. Walker [184].)

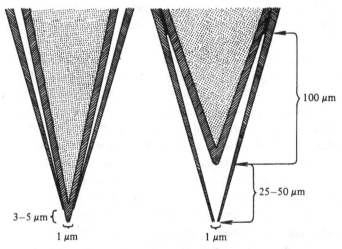

Fig. 4.6. The potassium microelectrode according to Walker [183]. (By permission of the American Chemical Society.)

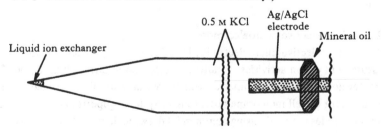

The chloride ion activity at the cell level was originally measured by micro-electrodes of the second kind, those with covered tips (see [85]); an ISM with a solid $AgCl/Hg_2S$ membrane has so far been used only for measurements in plant tissues [84].

A liquid-membrane ISM [66, 184, 185] was introduced into physiological laboratories by Walker [183]. It is schematically depicted in fig. 4.6. For reliable functioning, the internal surface of the tip must be siliconized over a distance of about $200\,\mu m$. The reaction,

$$\diagdown\!\!\!\!\underset{\diagup}{Si}\!-\!OH + R_3SiCl = -\underset{\diagup}{\overset{\diagdown}{Si}}\!-\!O\!-\!Si\!\overset{R}{\underset{R}{\diagdown}}\!\!\diagup R + HCl$$

takes place on the glass surface. As the siliconization agents, e.g. 5% trimethyl-chlorosilane in carbon tetrachloride [97], 2–4% tributylsilane in 1-chloro-naphthalene [89], 15% hexamethyldisilazane in pyridine [184] and various commercial mixtures are used.

When a simple ISM is used in a single cell, the simplest arrangement involves placing a reference electrode outside the cell and the measured electromotive voltage must be corrected for the membrane potential. The latter is measured using a micropipette inserted inside the cell and connected to another reference electrode. However, this procedure is unsatisfactory for excitable cells, inside which an electrical field is formed. Double-barrelled ISMs (fig. 4.7), introduced

Fig. 4.7. Various types of ion-sensitive microelectrodes: (*a* double-barrelled microelectrode; (*b*) coaxial double-barrelled microelectrode; (*c*) side-pore microelectrode. (After E. Ujec *et al.* [168] and P. Hník *et al.* [67].)

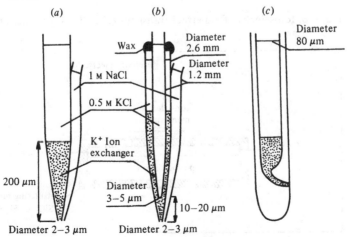

into electrophysiology by Coombs *et al.* [28], have been used successfully in such cases. In this system, only the inside of the ion-selective barrel can be siliconized, whereas the inside of the reference micropipette must remain hydrophilic. In the coaxial double-barrelled system [166, 168] (see fig. 4.7), the resistance of the ion-selective channel is substantially decreased. A robust side-pore ISM [67, 181] (see fig. 4.7) was constructed for measuring in muscles.

4.3   Ion-selective field-effect transistor ISFET

This is discussed in [74, 75, 163]. Attempts to combine an ion-selective electrode with the MOSFET (metal-oxide semiconductor field-effect transistor) began in 1970. According to Bergveld [12], who called this sensor an ISFET (ion-selective field-effect transistor), the effect of the electrical field in the MOSFET metal oxide is replaced by the ion-selective effect of the electrical double layer at the membrane/test solution interface. An intermediate between an ISE and an ISFET is a sensor with a membrane conductively connected with the MOSFET gate [78]. The construction of the ISFET is similar to that of the MOSFET, but the gate is replaced by an ion-selective membrane that is in contact with the solution [22, 76, 104, 115, 188, 189] (see fig. 4.8). In the insulator (SiO_2 or Si_3N_4) separating the membrane from the substrate (usually a p-type semiconductor) an electrical field is formed that increases or decreases the density of mobile charge carriers (holes) in the surface layer of the semiconductor. If the holes are repelled from the insulator/semiconductor interface into the semiconductor, a region of space charge is formed there. If the difference between the electrical potential inside the semiconductor and at the surface is sufficiently large, an excess of mobile electrons, an n-type conductive channel,

Fig. 4.8. Ion-selective field-effect transistor (ISFET). V_G denotes the gate voltage.

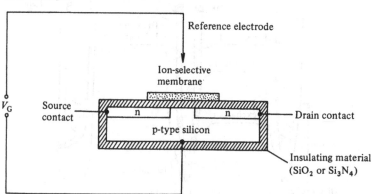

is formed at the surface, separated from the semiconductor interior by the space charge region. A further increase in the electrical field intensity no longer leads to a change in the potential difference between the surface and the inside of the semiconductor, but causes an increase in the electron density in the channel and thus also an increase in its conductance.

For correct function of the ISFET, a sufficiently large gate voltage, V_G, must be applied between the leads to the reference electrode and to the substrate, so that a sufficiently large potential difference is formed between the surface and the interior of the substrate for formation of the n-type conductive channel at the insulator/substrate interface. This channel conductively connects drain 1 and source 2, which are connected with the substrate by a p-n transition. On application of voltage V_D between the drain and the source, drain current I_D begins to pass. Under certain conditions the drain current is a linear function of the difference between V_G and the Volta potential difference between the substrate and the membrane.

Consider a liquid membrane that exhibits a Nernstian response for K^+ ions and is immersed in a solution with a K^+ ion activity of $a_{K^+}(5)$. A reference calomel electrode is connected with this solution by a saturated KCl liquid bridge. The whole system is depicted schematically in fig. 4.9. The gate voltage, V_G, is given by

$$V_G = \phi(9) - \phi(1) = (\bar{\mu}_e(1) - \bar{\mu}_e(9)/F), \qquad (4.1)$$

where $\bar{\mu}_e(1)$ and $\bar{\mu}_e(9)$ are the electrochemical potentials of the electrons in copper leads 1 and 9. The phase equilibria are

Fig. 4.9. The arrangement of contacting phases in an ISFET. V_G denotes the gate voltage.

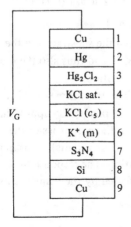

$$\bar{\mu}_e(1) = \bar{\mu}_e(2) = \mu_{Hg}^0 - \bar{\mu}_{Hg^+}(2) = \mu_{Hg}^0 - \bar{\mu}_{Hg^+}(3)$$
$$= \mu_{Hg}^0 - \mu_{Hg_2Cl_2}^0 + \bar{\mu}_{Cl^-}(3) = \mu_{Hg}^0 - \mu_{Hg_2Cl_2}^0 + \bar{\mu}_{Cl^-}(4)$$
$$= \mu_{Hg}^0 - \mu_{Hg_2Cl_2}^0 + \mu_{Cl^-}^0(wt) + RT \ln a_{Cl^-}(sat) - F\phi(4), \quad (4.2)$$

where μ_{Hg}^0 and $\mu_{Hg_2Cl_2}^0$ are the standard chemical potentials of Hg and Hg_2Cl_2, respectively, $\mu_{Cl^-}^0(wt)$ is the standard chemical potential of Cl^- in water and $a_{Cl^-}(sat)$ is the Cl^- activity in the saturated KCl solution. As the liquid-junction with KCl virtually eliminates the $\Delta\phi_L$ value between phases 4 and 5, it follows that

$$\phi(4) = \phi(5). \quad (4.3)$$

It further holds that

$$\bar{\mu}_{K^+}(5) = \bar{\mu}_{K^+}(6), \quad (4.4)$$

that is

$$\mu_{K^+}^0(wt) + RT \ln a_{K^+}(5) + F\phi(5) = \mu_{K^+}(m) + F\chi(m) + F\psi(6)$$
$$= \alpha_{K^+}(m) + \psi(6), \quad (4.5)$$

where $\mu_{K^+}^0(wt)$ is the standard chemical potential of K^+ in the aqueous solution, $\mu_{K^+}(m)$ is the chemical potential of K^+ in the membrane, $\chi(m)$ is the membrane surface potential and $\psi(6)$ is the outer membrane potential. The quantity, $\alpha_{K^+}(m) = \mu_{K^+}(m) + F\chi(m)$ is the real potential of ion K^+ in the membrane (the escape energy of K^+ from the membrane multiplied by -1, cf. [35a], chapter 3). Analogously it holds that

$$\bar{\mu}_e(8) = \alpha_e(s) - \psi(8) = \bar{\mu}_e(9), \quad (4.6)$$

where $\alpha_e(s)$ is the real potential of an electron in the substrate and $\psi(8)$ is the outer potential of the substrate. Combining (4.1), (4.2), (4.3), (4.5) and (4.6),

$$V_G - \psi(8) + \psi(6) = V_G - \Delta_m^s\psi$$
$$= (1/F)[\mu_{Hg}^0 - \mu_{Hg_2Cl_2}^0 + \mu_{KCl}^0(wt) + RT \ln a_{Cl^-}(sat)$$
$$- \alpha_{K^+}(m) - \alpha_e(s)] + (RT/F) \ln a_{K^+}(5), \quad (4.7)$$

where $\Delta_m^s\psi$ is the Volta potential difference between the substrate and the membrane. It can be seen that the quantity $V_G - \Delta_m^s\psi$ and thus also the drain current I_D, is a linear function of $\ln a_{K^+}$ in the test solution.

In contrast to potentiometry with ISEs, the drain current is measured with the ISFET and not the voltage. As the drain current depends only approximately linearly on $\Delta_m^s\psi$ and as the $\alpha_{K^+}(m)$ value depends on the properties of the membrane surface (for example, on the adsorption of surfactants), measurement of activities using an ISFET requires careful calibration. The response time depends on the membrane properties and is not affected by the components of the solid-phase sensor [162].

The hydrolysed surface of the Si_3N_4 insulator functions as a pH-sensitive membrane [90, 105, 116, 179]. A penicillin-sensitive ISFET is based on this membrane that is covered by an immobilized layer of penicillinase, converting penicillin into the penicillanic acid anion with liberation of hydrogen ions [24]. Another version of pH-sensitive ISFETs has membrane gates made of Ta_2O_5 [3] or of a suitable glass [39]. The latter ISFET with a gate made of alumino- or borosilicate glass is sensitive to sodium ions. Other ISFETs are sensitive to halide ions [22, 153, 178], K^+ [105, 115, 130] and Ca^{2+} [90, 105].

These miniaturized sensors are suitable for flow-injection analysis [130]. Similar systems have also been used in portable instruments [56]. A 'micro-ISFET' was constructed for intracellular determination of K^+ [53].

4.4   Composite systems

Composite potentiometric sensors involve systems based on ion-selective electrodes separated from the test solution by another membrane that either selectively separates a certain component of the analyte or modifies this component by a suitable reaction. This group includes gas probes, enzyme electrodes and other biosensors. Gas probes are discussed in this section and chapter 8 is devoted to potentiometric biosensors.

L. C. Clark first suggested in 1956 that the test solution be separated from an amperometric oxygen sensor by a hydrophobic porous membrane, permeable only for gases (for a review of the Clark electrode see [88]). The first potentiometric sensor of this type was the Severinghaus CO_2 electrode [150], with a glass electrode placed in a dilute solution of sodium hydrogenocarbonate as the internal sensor (see fig. 4.10). As an equilibrium pressure of CO_2, corresponding to the CO_2 concentration in the test solution, is established in the

Fig. 4.10. A potentiometric gas probe (manufactured by Orion Research).

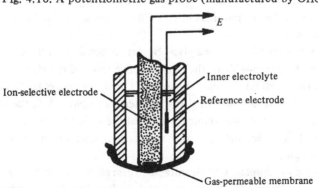

membrane pores, the same CO_2 concentration is also attained in the internal solution at the glass electrode. The pH of this solution, sensed by the glass electrode, is

$$pH = \log \frac{a_{HCO_3^-}}{K_1 a_{H_2CO_3}} = \frac{a_{HCO_3^-}}{K_1 K_s p_{CO_2}}, \qquad (4.8)$$

where K_1 is the first dissociation constant of carbonic acid and K_s is the constant for the equilibrium between gaseous CO_2 and carbonic acid.

Therefore, the ISE potential depends on the CO_2 partial pressure with Nernstian slope. Contemporary microporous hydrophobic membranes permitted the construction of a number of gas probes, developed mainly by the Orion Research Company (for a survey see [143]. The most important among these sensors is the ammonia electrode, in which ammonia diffusing through the membrane affects the pH at a glass electrode. Other electrodes based on similar principles respond to SO_2, HCN, H_2S (with an internal S^{2-} ISE), etc. The ammonia probe has a better detection limit than the ammonium ion ISE based on the non-actin ionophore. The response time of gas probes depends mostly on the rate of diffusion of the test gas through the microporous medium [77, 143].

4.5   Calibration of ion-selective electrodes

As mentioned above, the ISE potential depends directly on the test ion activity and only indirectly on its concentration. At a single concentration, the individual ion activities can vary considerably in their dependence on the solution composition.

It need not be emphasized that the activity of free ions strongly decreases during complex formation. This effect can readily be taken into account when the kind and concentrations of the complexing agents is known, but in complicated media, such as body fluids, the complexation usually cannot be described using thermodynamic calculations.

In samples with variable ionic compositions, the liquid-junction potential can also vary considerably and these effects must be considered in the methods for ISE calibration.

The IUPAC Commission for Analytical Nomenclature defines the calibration curve [138] as the dependence of the electromotive force of the given ISE – reference electrode cell on the logarithm of the activity or concentration of the given substance. It is recommended that the potential be plotted on the ordinate (the vertical axis) and the logarithmic function of the activity or concentration on the abscissa (the horizontal axis), with the concentration increasing from the left to the right.

The calibration is simplest when a suitable excess electrolyte is added to the test solution to maintain the ionic strength and the liquid–junction potential constant and sometimes also to decompose complexes present in the test solution. A typical example of such an electrolyte is TISAB, used in the determination of fluoride ions (see section 6.4). The calibration curve in EMF–determinand concentration coordinates can be used directly for the determination of the test ion concentration from the measured electromotive force.

If the analyte cannot be pretreated in the way described above, methods must be found for the use of (3.1.5) or (3.1.7) even when activity coefficients and the liquid–junction potential vary with variations in the test solution composition.

At least a partial solution to this problem is attained by the conventional activity scale method [5, 6, 7, 9, 10, 11]. This procedure was first used by Bates and Guggenheim [8] when formulating the operational definition of pH (see [86a], chapter 1), on the basis of which the National Bureau of Standards in the USA developed a method for determining conventional hydrogen ion activities. The basic assumption is the use of the Debye–Hückel relationship for the individual activity of chloride ions:

$$- \log a_{Cl^-} = AI^{1/2}/(1 + 1.5I^{1/2}), \tag{4.9}$$

where $A = 0.512 \text{ mol}^{-1/2} \text{ kg}^{1/2}$ at 25 °C and I is the ionic strength (on a molality scale). This assumption was also used by Bates and Alfenaar [7] in calculating the individual ion activities in solutions suitable for ISE calibration. As the activities of individual ions and mean activities of salts must be consistent, i.e. it must hold for salt $A_n B_m$ (see e.g. [86a], chapter 1) that

$$(a_{A^{m+}}^n a_{B^{n-}}^m)^{1/(m+n)} = a_{\pm}(A_n B_m), \tag{4.10}$$

it is easy, at least at low ionic strength values, to calculate the individual ion activities from the known mean activities. First, the cation activities are calculated from the mean activities of their chlorides, using (4.9). Then the activities of other anions are calculated in the same way from the mean activities of their salts with the cations whose individual activity coefficients have already been determined.

Of course, (4.9) cannot be used at higher ionic strengths. A more precise procedure is then required, such as the Robinson-Stokes equation for the mean molal activity coefficient [11] (see also [86a], chapter 1)

$$\ln \gamma_{\pm} = |z_+ z_-| \ln f_{DH} - h/\nu \ln a_{wt} - \ln [1 + 0.018(\nu - h)], \tag{4.11}$$

where z_+ and z_- are the charge numbers of the cation and anion, respectively, h is the hydration number of the electrolyte as a whole (for the newest h scale see [10]), ν is the number of ions formed by dissociation of one molecule of the electrolyte, a_{wt} is the solvent (water) activity and the Debye–Hückel term, f_{DH}, is

$$\ln f_{DH} = -AI^{1/2}/(1 + BaI^{1/2}),$$

where A and B are the constants in the Debye-Hückel theory and a is the effective radius of the ion ([86a], chapter 1). This equation can be used up to molalities $12/h$. Assuming that the chloride ion is unhydrated, the simple relationship [11],

$$\log \gamma_+ = \log \gamma_\pm + 0.00782 \, hm\phi, \tag{4.12}$$

follows from (4.9) and (4.10), where m is the solution molality and ϕ is the electrolyte osmotic coefficient (see [86a], chapter 1).

The individual activity coefficients calculated from (4.12), suitable for calibration of ISEs for chloride ions, the alkali metal and alkaline earth ions, are given in tables 4.1 and 4.2. Ion activity scales have also been proposed for KF [141], choline chloride [98], for mixtures of electrolytes simulating the composition of the serum and other biological fluids (at 37 °C) [106, 107], for alkali metal chlorides in solutions of bovine serum albumine [132] and for mixtures of electrolytes analogous to sea water [140].

The calibration of ISEs using the tabulated activity values is especially simple when the dependence of the ISE potential on the determinand activity is Nernstian (3.1.5). If the liquid–junction potential is negligible or constant, the determinand activity $a_{J^+}(X)$ can be found, using the standard activity $a_{J^+}(S)$, from the relationship [7],

$$\log a_{J^+}(X) = \log a_{J^+}(S) + (E_X - E_S)zF/2.303 \, RT, \tag{4.13}$$

where E_X is the electromotive force of cell (3.1.1) with the determinand in solution 1 and E_S is that with the standard in solution 1. The standard activity is chosen as close as possible to the determinand activity.

When the condition of negligible or constant liquid–junction potential is not satisfied, a correction must be made for the change in this potential,

$$\delta\Delta\phi_L = \Delta\phi_L(X) - \Delta\phi_L(S) \tag{4.14}$$

where $\Delta\phi_L(X)$ and $\Delta\phi_L(S)$ are the liquid–junction potentials calculated from the Henderson equation (2.6.12). Equation (4.13) must then be rearranged to the form,

$$\log a_{J^+}(X) = \log a_{J^+}(S) + (E_X - E_S - \delta\Delta\phi_L)zF/2.303 \, RT. \tag{4.15}$$

The electrode calibration is subject to complications connected with calibration solutions with low standard concentrations. For example, in calibration of the Ag_2S ISE with $AgNO_3$ solutions, the lowest practicably attainable concentration is 4×10^{-7} M when a Teflon vessel is used. Teflon exhibits the lowest adsorption of ions on the vessel walls; the adsorption increases in the order [35],

Teflon < Vycor < polyethylene < dried Pyrex < Pyrex.

On the other hand, complexes of silver can be used for calibration down to the silver ion activity, 10^{-24} M [142, 176].

Table 4.1. *Single ion activity coefficients (molal scale) for uni-univalent chlorides at 25 °C derived from hydration theory* [11].

m	$\gamma+$	$\gamma-$	$\gamma+$	$\gamma-$
	HCl		LiCl	
0.1	0.807	0.785	0.799	0.781
0.2	0.788	0.746	0.775	0.739
0.5	0.812	0.706	0.786	0.695
1.0	0.940	0.697	0.882	0.680
2.0	1.421	0.717	1.233	0.688
3.0	2.357	0.735	1.893	0.706
	NaCl		KCl	
0.1	0.783	0.773	0.773	0.768
0.2	0.744	0.726	0.722	0.714
0.5	0.701	0.661	0.659	0.639
1.0	0.697	0.620	0.623	0.586
2.0	0.756	0.590	0.610	0.538
3.0	0.870	0.586	0.626	0.517
4.0	1.038	0.591	0.659	0.506
5.0	1.272	0.600		
6.0	1.594	0.610		
	RbCl		CsCl	
0.1	0.766	0.762	0.756	0.756
0.2	0.712	0.706	0.694	0.694
0.5	0.640	0.628	0.606	0.606
1.0	0.594	0.572	0.544	0.544
2.0	0.568	0.525	0.496	0.496
3.0	0.569	0.505	0.479	0.479
4.0	0.584	0.496	0.474	0.474
5.0	0.606	0.492	0.475	0.475
	NH₄Cl			
0.1	0.772	0.768		
0.2	0.722	0.714		
0.5	0.657	0.641		
1.0	0.619	0.588		
2.0	0.601	0.541		
3.0	0.608	0.518		
4.0	0.624	0.502		
5.0	0.645	0.490		
6.0	0.667	0.477		

Systems containing a certain concentration of a complex in solution, preferably with a metal-to-ligand ratio of 1:1, common for polydentate ligands, and an excess of a complexing agent with the properties of a Brønsted acid, act as metal buffers [158]. These systems were used for ISE calibration [15, 45, 55, 144], the negative log of the activity of the free metal ion, pM, being given by

$$pM = \log K_{ML} + \log c_L - \log c_M + \log F \qquad (4.16)$$

where K_{ML} is the stability constant of complex ML, c_L and c_M are the overall concentrations of ligand L and metal M, respectively (c_M virtually equals the concentration of complex ML) and F is a function of the pH and the activity coefficients of all the substances participating in the formation of complex ML. The activity of a free metal ion can be varied over a wide range by varying the pH. Analogously, by changing the pH in solutions containing the system, $S^{2-} \rightleftharpoons HS^- \rightleftharpoons H_2S$, the sulphide ion activity can be decreased down to a value of 10^{-19} mol dm^{-3} and thus the Ag_2S ISE potential can be affected [68]. It should be borne in mind that ISE calibration with metal buffers is unsuitable when the electrode detection limit is given by the solubility of the membrane ion-exchanger systems. The Nernstian calibration curve beyond the detection limit then gives erroneous information on the ISE behaviour in real test solutions.

Table 4.2. *Single ion activity coefficients (molal scale) for alkaline earth chlorides at 25 °C derived from hydration theory* [11].

m	γ_{2+}	γ_-	γ_{2+}	γ_-
	MgCl$_2$		CaCl$_2$	
0.0333			0.378	0.784
0.1	0.279	0.726	0.269	0.719
0.2	0.239	0.697	0.224	0.685
0.5	0.234	0.688	0.204	0.665
1.0	0.344	0.732	0.263	0.690
2.0	1.439	0.898	0.768	0.804
	SrCl$_2$		BaCl$_2$	
0.1	0.266	0.717	0.259	0.712
0.2	0.218	0.681	0.204	0.668
0.5	0.190	0.653	0.165	0.630
1.0	0.226	0.667	0.167	0.620
1.8			0.229	0.642
2.0	0.542	0.753		

4.6   Selectivity coefficient

The selectivity coefficient was defined in chapter 3 and several theoretical relationships were given for this quantity for various ISE systems. Several methods have been proposed [38, 120, 123, 135] for the determination of selectivity coefficients; two basic methods were recommended by the IUPAC Commission for Analytical Nomenclature [138].

Method 1 is the constant interferent concentration method (mixed solution technique) [123]. The EMF of a cell consisting of an ISE and a reference electrode is measured at constant interferent activity a_K and variable determinand activity a_J. The EMF values are plotted against the logarithm of the determinand activity. The intercept of the asymptotes to this curve (cf. chapter 3) gives the a_J value that is used for the calculation of $k_{J,K}^{pot}$ from the relationship,

$$k_{J,K}^{pot} = \frac{a_J}{a_K^{z_J/z_K}}. \tag{4.17}$$

Method 2 is the separate solution method [38, 135]. The EMF of a cell consisting of an ISE and a reference electrode is measured in a solution of determinand J alone (in the absence of interferent K) and then in a solution of interferent K alone (in the absence of determinand J), the activities of the determinand and the interferent in the two solutions being the same. If the respective values of EMF are E_1 and E_2 and E_{ISE} for both ions are given by the simple Nernst equation, then $k_{J,K}^{pot}$ can be calculated from the relationship,

$$\log k_{J,K}^{pot} = \frac{E_2 - E_1}{2.303\, RT/z_J F} + \left(1 + \frac{z_J}{z_K}\right) \log a_J \tag{4.18}$$

The latter method should only be used when the constant interferent concentration method is unsuitable or impracticable.

Buck [20] introduced the concept of the apparent selectivity coefficient, $k_{J,K}^{pot}(app)$, defined by the relationship,

$$\ln k_{J,K}^{pot}(app) = \ln\{\exp[(E_{JK} - E_J)/S] - 1\} - \ln a_K/a_J \tag{4.19}$$

where E_{JK} is the ISE potential in the presence of ions J and K at activities a_J and a_K, E_J is the ISE potential in the presence of ion J alone at activity a_J and S is the slope of the E_{ISE} versus a_J dependence,

$$E = E_0 + S \ln a_J \tag{4.20}$$

Equation (4.19) can be used only when (4.20) is valid. In simple systems (rapid processes at the membrane/electrolyte interface and a simple diffusion potential in the membrane) the apparent selectivity coefficient is a function of the a_J/a_K ratio alone, whereas in more complicated systems it also depends on the activities of J and K.

The Appendix lists selectivity coefficients, specifying the methods of their determination.

4.7 Response time

If an ISE is transferred from a solution with determinand activity a_1 to a solution with determinand activity a_2, the E_{ISE} value does not change instantaneously from initial value E_1 to the value corresponding to activity a_2, E_2, but the time course of E_{ISE} is given by the curve shown in fig. 4.11. Practical response time τ_{90} has been defined by the IUPAC Commission for Analytical Nomenclature [138] as the time during which E_{ISE} changes from value E_1 to value $E_1 + 0.9(E_2 - E_1)$, i.e. during which E_{ISE} changes by 90% of the total change from E_1 to E_2. When reporting data concerning the response time, the solution composition, solution stirring, electrode pretreatment and the temperature should be specified.

The simplest measurement of the response time of macroelectrodes involves the determination of the time elapsed from the transfer of the ISE from the solution in which it was stored into the test solution. Data obtained in this way are only significant for response times of at least tens of seconds. For short response times, flow-through [44, 131] or injection [92] methods are suitable. For microelectrodes, a rapid immersion method [181], a flow-through [166] or an iontophoretic method [97] have given good results. In the last method

Fig. 4.11. Response time τ_{90} and time of relaxation τ (4.27).

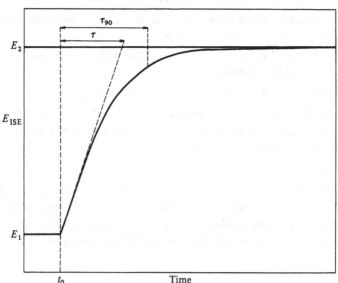

a constant current pulse is applied to a KCl solution in a glass capillary with orifice close to a K^+-sensitive microelectrode. In this way a defined, time-dependent concentration distribution is formed, reaching to the electrode surroundings. This method permits the measurement of response times below 1 ms.

As pointed out by Shatkay [152], there was no theory of the response time before 1975, except for the work of Markovic and Osborn [99], and the time dependence of E_{ISE} on a change in the concentration was characterized by interpolation formulae, such as the exponential,

$$E(t) = E_1 + (E_2 - E_1) \exp(-kt) \tag{4.21}$$

or the hyperbolic relationship

$$E(t) = E_1 + (E_2 - E_1)kt/(1 + kt). \tag{4.22}$$

The response time is affected by the following factors [2, 186]:

1. The time constant of the measuring instrument [166, 167].
2. Determinand diffusion through the stagnant layer [99].
3. The rate of the charge-transfer reaction across the membrane/solution interface, leading to charging of the electrical double layer at this interface [87].
4. The rate of the exchange reaction between the determinand in the membrane and an interferent in the test solution.
5. Determinand diffusion into the membrane [114].
6. Establishment of the diffusion potential across the membrane [80].
7. Dissolution of the membrane active component in the test solution [13].

Fig. 4.12. Distribution of determinand concentration in the neighbourhood of ISE.

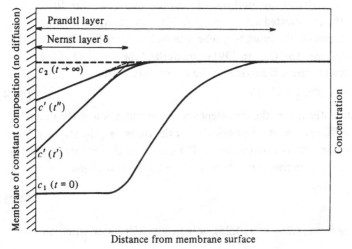

The first factor, especially important with ion-selective microelectrodes, can be eliminated by a suitable modification of the measuring instrument, notably by the use of a coaxial microelectrode (see [167] and section 4.2). If an interferent is present in the solution at a concentration at which it does not affect the ISE potential, factors 4 and 6 are not operative. Penetration of the determinand into the membrane, factor 5, is very important for the response times of ISEs with ionophores in their membranes, provided that no hydrophobic anion is present in the membrane solution, as has been theoretically treated by Morf *et al.* [114]. As shown in section 3.3, the presence of a hydrophobic anion stabilizes the conditions in the membrane, with a marked effect on the shortening of the response time [93].

When considering a membrane with a stable composition and a determinand concentration sufficiently higher than the detection limit, so that factor 7 is unimportant, the response time depends on factors 2 and 3.

As follows from the hydrodynamic properties of systems involving phase boundaries (see e.g. [86a], chapter 2), the hydrodynamic, Prandtl or stagnant layer is formed during liquid movement along a boundary with a solid phase, i.e. also at the surface of an ISE with a solid or plastic membrane. The liquid velocity rapidly decreases in this layer as a result of viscosity forces. Very close to the interface, the liquid velocity decreases to such an extent that the material is virtually transported by diffusion alone in the Nernst layer (see fig. 4.13). It follows from the theory of diffusion transport toward a plane with characteristic length l, along which a liquid flows at velocity V_0, that the Nernst layer thickness, δ, is given approximately by the expression,

$$\delta \approx (\eta l D^2 / \rho V_0^3)^{1/6}, \qquad (4.23)$$

where η is the viscosity of the solution, ρ is its density and D is the diffusion coefficient of the transported substance. With ISEs, the diameter of the electrode tip, including the insulating tube, can be considered as characteristic dimension l. Markovic and Osborn [99] characterized the flux $J(t)$ of the determinand toward the ISE surface by the Nernst relationship,

$$J(t) = -(D(c_2 - c')/\delta), \qquad (4.24)$$

where c_2 is the initial and c' the time-dependent concentration of the determinand at the ISE surface. At the beginning of the transport, the stagnant layer contains a solution with concentration c_1. The change in the average concentration, $c_{av} = \frac{1}{2}(c_2 - c')$, in the Nernst layer is given by the expression,

$$\frac{d\,c_{av}}{dt} = \frac{J(t)}{\delta} \qquad (4.25)$$

On substituting from (4.24) into (4.25), solving the differential equation and

substituting the resultant expression for c' into (3.15), the relationship for
$E(t)$ is obtained, in the case of a monovalent determinand

$$E(t) = E_2 + \frac{RT}{F} \ln \left[1 - \left(1 - \frac{c_1}{c_2} \right) \exp \left(-\frac{t}{\tau} \right) \right] \qquad (4.26)$$

where τ is the relaxation time that is lower than τ_{90}, but is also a suitable
characteristic of the response rate,

$$\tau = \frac{\delta^2}{2D} \approx \frac{l}{V_0} \left(\frac{\eta \rho}{D} \right)^{1/3} . \qquad (4.27)$$

At stirring rates of 10 cm s^{-1} and with common values of l, η, δ and D for
aqueous solutions, the relaxation times obtained are of the order of seconds,
corresponding to the reality.

On a change in the ISE potential between the values of E_1 and E_2, the
electrical double layer at the membrane/test solution interface must accept
a sufficient electrical change, corresponding to the ISE potential, E_2. The only
way of receiving this charge in the zero-current state is the transfer of the
determinand ions across the interface. The current corresponding to the charge-
transfer (2.5.5), j_t, must equal the charging current,

$$j_c = C \frac{d\Delta V}{dt} , \qquad (4.28)$$

where C is the differential capacity of the membrane/analyte interface. In the
case of an ion-selective microelectrode, then the effect of the transport
toward the membrane can be neglected in view of natural convection and the
solution of a differential equation based on (2.5.5) and (4.8) yields a real
response time that is of the order of milliseconds [87], in agreement with the
experimental values [167].

4.8   Temperature coefficient, drift and life-time
The temperature coefficient of the ISE potential has received relatively
little attention. As follows from (3.1.7), the constant term (the ISE 'standard'
potential), the determinand and interferent activity coefficients and the selectivity coefficient, liquid–junction potentials and, of course, also the $RT/z_i F$
coefficient, depend on the temperature [118]. When the internal reference
electrode and the reference electrode in the test solution are identical, the
interferent activity sufficiently low and the liquid–junction potentials negligible,
then the constant term depends on the determinand activity in the electrode
internal solution alone and thus the temperature coefficient of the measured
EMV depends only on the temperature coefficient of the determinand activity
coefficient and on the $RT/z_i F$ coefficient. Measuring instruments are usually

provided with automatic compensation for this effect. The temperature coefficient of the selectivity coefficient is a function of the standard enthalpy of the corresponding exchange reaction and, when a diffusion potential is formed across the membrane, also of the transport activation energies of the participating ions.

The ISE potential drift is connected with changes in the ISE surface structure and with dissolution of ion-exchanger ions or ionophores in test solutions. The values given in the literature differ considerably and may be as large as 2 mV per day.

The ISE life-time is closely connected with the drift and is at least one year for good electrodes. With some systems, e.g. enzyme electrodes (see chapter 8), the life-time is only a few weeks. It follows from the results of Oesch and Simon [119] that the life-time of electrodes based on ionophores in solvent-polymeric membranes depends on the kinetics of dissolution of the ionophore and the plasticiser in the analyte. If both the ionophore and the membrane solvent have distribution coefficients between water and the membrane greater than 10^6, then the ISE life-time is at least one year.

4.9   Ion-selective electrodes in non-aqueous media

The use of ISEs in non-aqueous media (for a survey see [125, 128]) is limited to electrodes with solid or glassy membranes. Even here there are further limitations connected with membrane material dissolution as a result of complexation by the solvent and damage to the membrane matrix or to the cement between the membrane and the electrode body. Silver halide electrodes have been used in methanol, ethanol, *n*-propanol, *iso*-propanol and other aliphatic alcohols, dimethylformamide, acetic acid and mixtures with water [40, 81, 121, 128]. The slope of the ISE potential dependence on the logarithm of the activity decreases with decreasing dielectric constant of the medium. With the fluoride ISE, the theoretical slope was found in ethanol–water mixtures [95] and in dimethylsulphoxide [23], and with PbS ISE in alcohols, their mixtures with water, dioxan and dimethylsulphoxide [134]. The standard Gibbs energies for the transfer of ions from water into these media were also determined [27, 30] using ISEs in non-aqueous media.

References for Chapter 4

1 R. N. Adams, *Electrochemistry at Solid Electrodes*, M. Dekker, New York (1969), p. 280.
2 T. Akiyama, K. Kinoshita, Y. Horita and E. Niki, *Nippon Kagaku Kaishi*, 1431 (1980).
3 T. Akiyama, T. Sugano and E. Niki, *Bunseki Kagaku* **29**, 584 (1980).
4 A. Ansaldi and S. I. Epstein, *Anal. Chem.* **45**, 595 (1973).

5 R. G. Bates, *Determination of pH, Theory and Practice*, 2nd edition, John Wiley & Sons, New York (1973).

6 R. G. Bates, *Pure Appl. Chem.* **36**, 407 (1973).

7 R. G. Bates and M. Alfenaar, Activity Standards for Ion-Selective Electrodes, Chapter 6, *Ion-Selective Electrodes* (ed. R. A. Durst), NBS Special Publication No. 317, Washington (1969).

8 R. G. Bates and E. A. Guggenheim, *Pure Appl. Chem.* **1**, 163 (1960).

9 R. G. Bates and R. A. Robinson, *Pure Appl. Chem.* **37**, 575 (1974).

10 R. G. Bates and R. A. Robinson, Trends in the standardization of ion-selective electrodes, *Ion-Selective Electrodes* (ed. E. Pungor and I. Buzás), Conference 1977, Akadémiai Kiadó, Budapest (1978), p. 3.

11 R. G. Bates, B. R. Staples and R. A. Robinson, *Anal. Chem.* **42**, 867 (1970).

12 P. Bergveld, *IEEE Trans. Biomed. Eng.* **19**, 342 (1972).

13 W. J. Blaedel and D. E. Dinwiddie, *Anal. Chem.* **46**, 873 (1974).

14 R. Bloch, A. Shatkay and H. A. Saroff, *Biophys. J.* **7**, 865 (1967).

15 R. Blum and H. M. Fog, *J. Electroanal. Chem.* **34**, 485 (1972).

16 J. M. Robbitt, J. F. Colaruotolo and S. J. Huang, *J. Electrochem. Soc.* **120**, 773 (1973).

17 I. A. Borisova and I. A. Gurev, *Zavod. Lab.* **45**, 309 (1979).

17a N. Bottazzini and V. Crespi, *Chim. Ind. (Milan)* **52**, 866 (1970).

18 H. M. Brown and J. D. Owen, *Ion-Sel. El. Rev.* **1**, 145 (1979).

19 E. B. Buchanan and J. L. Seago, *Anal. Chem.* **40**, 517 (1968).

20 R. P. Buck, *Anal. Chim. Acta* **73**, 321 (1974).

21 R. P. Buck, Theory and principles of membrane electrodes, Chapter 1 of *Ion-Selective Electrodes in Analytical Chemistry* (ed. H. Freiser), Vol. I, Plenum Press, New York (1978).

22 R. P. Buck and D. E. Hackleman, *Anal. Chem.* **49**, 2315 (1977).

23 J. N. Butler, Thermodynamic studies, Chapter 5 of *Ion-Selective Electrodes* (ed. R. A. Durst), NBS Special Publication No. 317, Washington (1969).

24 S. Caras and J. Janata, *Anal. Chem.* **52**, 1935 (1980).

25 W. R. Cattral, Heterogeneous-membrane, carbon-supported and coated-wire ion-selective electrodes, in *Ion-Selective Electrode Methodology*, Vol. I (ed. A. K. Covington), CRC Press, Boca Raton (1979), p. 131.

25a R. W. Cattral and K. T. Fong, *Talanta* **25**, 541 (1978).

26 R. W. Cattral and H. Freiser, *Anal. Chem.* **43**, 1905 (1971).

26a S. N. K. Chaudhari and K. L. Cheng, *Mikrochim. Acta*, 411 (1979).

26b *Chemically Sensitive Electronic Devices, Principles and Applications* (ed. J. Zemel and P. Bergveld), Elsevier Sequoia, Lausanne (1981).

27 C. J. Coetzee and W. K. Istone, *Anal. Chem.* **52**, 53 (1980).

28 J. S. Coombs, J. C. Eccles and P. Fatt, *J. Physiol. (London)* **130**, 291 (1955).

29 A. K. Covington, Heterogeneous membrane electrodes, Chapter 3 of *Ion-Selective Electrodes* (ed. R. A. Durst), NBS Special Publication No. 317, Washington (1969).

30 A. K. Covington and J. M. Thain, *J. Chem. Soc. Faraday Trans.* **71**, 78 (1975).

31 A. Craggs, L. Keil, G. J. Moody and J. D. R. Thomas, *Talanta* **22**, 907 (1975).

32 A. Craggs, G. J. Moody and J. D. R. Thomas, *J. Chem. Educ.* **51**, 541 (1974).

33 J. E. W. Davies, G. J. Moody, W. M. Prince and J. D. R. Thomas, *Lab. Pract.* **1973**, 20.

34 J. E. W. Davies, G. J. Moody and J. D. R. Thomas, *Analyst* **97**, 87 (1972).

35 R. A. Durst and B. T. Duhart, *Anal. Chem.* **42**, 1002 (1970).

36 L. Ebdon, A. T. Ellis and G. C. Corfield, *Analyst* **104**, 730 (1979).

37 A. T. Ellis, G. C. Corfield and L. Ebdon, *Anal. Proc. (London)* 17, 48 (1980).
38 G. Eisenman, D. O. Rudin and J. U. Casby, *Science* 126, 871 (1957).
39 M. Esashi and T. Matsuo, *IEEE Trans. Biomed. Eng.* 25, 184 (1978).
40 W. H. Ficklin and W. C. Gotschall, *Anal. Lett.* 6, 317 (1973).
41 U. Fiedler and J. Růžička, *Anal. Chim. Acta* 67, 179 (1973).
42 T. A. Fjeldly and K. Nagy, *J. Electrochem. Soc.* 127, 1299 (1980).
43 B. Fleet, G. P. Bound and D. R. Sandbach, *Bioelectrochem. Bioenerg.* 3, 158 (1976).
44 B. Fleet, T. H. Ryan and M. J. Brand, *Anal. Chem.* 46, 12 (1974).
45 M. S. Frant and J. W. Ross, *Science* 154, 1553 (1966).
46 H. Freiser, *Res./Dev.* 27, 28, 32 (1976).
47 H. Freiser, Coated wire ion-selective electrodes, Chapter 2 of *Ion-Selective Electrodes in Analytical Chemistry* (ed. H. Freiser) Vol. 2, Plenum Press, New York (1980).
48 *Glass Electrodes for Hydrogen and Other Cations. Principles and Practice* (ed. G. Eisenman), M. Dekker, New York (1967).
49 G. H. Griffiths, G. J. Moody and J. D. R. Thomas, *Analyst* 97, 420 (1972).
50 I. A. Gurev, E. A. Gushina and E. N. Mitina, *Zh. Anal. Khim.* 34, 1184 (1979).
51 I. A. Gurev, G. M. Lizanova and N. S. Bulanova, *Zh. Anal. Khim.* 34, 1809 (1979).
52 I. A. Gurev and T. S. Vyatchanina, *Zh. Anal. Khim.* 34, 976 (1979).
53 A. Haemmerli, J. Janata and H. M. Brown, *Anal. Chem.* 52, 1179 (1980).
54 E. H. Hansen, C. G. Lamm and J. Růžička, *Anal. Chim. Acta* 59, 403 (1972).
55 E. H. Hansen and J. Růžička, *Anal. Chim. Acta* 72, 365 (1974).
56 J. Harrow, J. Janata, R. L. Stephen and W. J. Kolff, *Proc. EDTA* 17, 179 (1980).
57 G. Heidecke, J. Kropf, G. Stock and J. G. Schindler, *Fresenius Z. Anal. Chem.* 303, 364 (1980).
58 K. Hiiro, G. J. Moody and J. D. R. Thomas, *Talanta* 22, 918 (1975).
59 J. A. M. Hinke, Cation-selective microelectrodes for intracellular use, Chapter 17 of *Glass Electrodes for Hydrogen and Other Cations* (ed. G. Eisenman), M. Dekker, New York (1969).
60 H. Hirata, M. Axai and N. Toonooka, *Nippon Kagaku Kaishi* 10, 1475 (1980).
61 H. Hirata and K. Date, *Anal. Chim. Acta* 51, 209 (1970).
62 H. Hirata and K. Date, *Talanta* 17, 838 (1970).
63 H. Hirata and K. Higashiyama, *Anal. Chim. Acta* 54, 415 (1971).
64 H. Hirata and K. Higashiyama, *Anal. Chim. Acta* 57, 476 (1971).
65 H. Hirata and K. Higashiyama, *Talanta* 19, 391 (1972).
66 P. Hník, E. Syková, N. Kříž and F. Vyskočil: Determination of ion activity changes in excitable tissues with ion-selective microelectrodes, Chapter 5 of *Medical and Biological Applications of Electrochemical Devices* (ed. J. Koryta), John Wiley and Sons, Chichester (1980).
67 P. Hník, F. Vyskočil, N. Kříž and M. Holas, *Brain Res.* 40, 559 (1972).
68 T. M. Hsen and G. A. Rechnitz, *Anal. Chem.* 40, 1054 (1968).
69 A. Hulanicki and M. Trojanowicz, *Anal. Chim. Acta* 87, 411 (1976).
70 *Ion-Selective Microelectrodes* (ed. H. J. Berman and N. C. Hebert), Plenum Press, New York (1974).
71 *Ion-Selective Microelectrodes and Their Use in Excitable Tissues* (ed. E. Syková, P. Hník and L. Vyklický), Plenum Press, New York (1981).
72 Israel Patent No. 21709 (1966).
73 H. J. James, G. P. Garmack and H. Fresier, *Anal. Chem.* 44, 856 (1972).
74 J. Janata and R. J. Huber, Chemically sensitive field effect transistors, Chapter 3 of

Ion-Selective Electrodes in Analytical Chemistry, Vol. 2 (ed. H. Fresier), Plenum Press, New York (1980).
75 J. Janata and R. J. Huber, *Ion-Sel. El. Rev.* 1, 31 (1979).
76 J. Janata and S. D. Moss, *Biomed. Eng.* 11, 241 (1976).
77 M. A. Jensen and G. A. Rechnitz, *Anal. Chem.* 51, 1972 (1979).
78 J. S. Johannessen, T. A. Fjeldly and K. Nagy, *Phys. Scripta* 18, 464 (1978).
79 A. Jyo, T. Imato, K. Fukamachi and N. Ishibashi, *Chem. Lett.* 7, 815 (1977).
80 B. Karlberg, *J. Electroanal. Chem.* 42, 115 (1973).
81 N. A. Kazaryan, A. P. Kreshkov and T. M. Syrykh, *Zh. Fiz. Khim.* 47, 2590 (1973).
82 O. Kedem, E. Loebel and M. Furmansky, *Ger. Offen* 2,027,127 (1970).
83. L. Keil, G. J. Moody and J. D. R. Thomas, *Analyst* 102, 274 (1977).
84. L. S. Kelday, D. J. F. Bowling and M. G. Penny, *J. Exp. Bot.* 28, 31 (1977).
85 R. D. Keynes, *J. Physiol. (London)* 169, 690 (1963).
86 I. M. Kolthoff and H. L. Sanders, *J. Am. Chem. Soc.* 59, 416 (1937).
86a J. Koryta, J. Dvořák and V. Boháčková, *Electrochemistry*, Methuen & Co., London (1970).
87 J. Koryta, Theory of ion-selective electrodes, in *Ion-Selective Microelectrodes and Their Use in Excitable Tissues* (ed. E. Syková, P. Hník and L. Vyklický), Plenum Press, New York (1981), p. 3.
88 F. Kreuzer, H. P. Kimmich and M. Březina, Polarographic determination of oxygen in biological materials, Chapter 6 of *Medical and Biological Applications of Electrochemical Devices* (ed. J. Koryta), John Wiley & Sons, Chichester (1980).
89 D. L. Kunze, *Circ. Res.* 41, 122 (1977).
90 I. R. Lanks and J. M. Zemel, *IEEE Trans. Electron. Devices* 26, 1959 (1979).
91 A. Liberti, Ion-selective electrodes from heterogeneous polythene membranes, in *Ion-Selective Electrodes* (ed. E. Pungor), Symposium 1972, Akadémiai Kiadó, Budapest (1973), p. 37.
92 E. Lindner, K. Tóth, E. Pungor, W. E. Morf and W. Simon, *Anal. Chem.* 50, 1627 (1978).
93 E. Lindner, K. Tóth and E. Pungor, *Anal. Chem.* 48, 1071 (1976).
94 G. Ling and R. W. Gerard, *J. Cellular Comp. Physiol.* 34, 382 (1949).
95 J. J. Lingane, *Anal. Chem.* 40, 935 (1968).
96 O. O. Lyalin and M. S. Turaeva, *Zh. Anal. Khim.* 31, 1879 (1976).
97 H. D. Lux and E. Neher, *Exper. Brain Res.* 17, 190 (1973).
98 J. B. Macaskill, M. S. Mohan and R. G. Bates, *Anal. Chem.* 49, 209 (1977).
99 P. L. Markovic and J. O. Osborn, *AIChE.* 19, 504 (1973).
100 C. R. Martin and H. Freiser, *J. Chem. Educ.* 57, 512 (1980).
101 A. Marton and E. Pungor, *Anal. Chim. Acta* 54, 209 (1971).
102 M. Mascini and A. Liberti, *Anal. Chim. Acta* 47, 339 (1969).
103 M. Mascini and A. Liberti, *Anal. Chim. Acta* 51, 231 (1970).
104 T. Matsuo and K. D. Wise, *IEEE Trans. Biomed. Eng.* 21, 485 (1974).
105 P. T. McBride, J. Janata, P. A. Comte, S. D. Moss and C. C. Johnson, *Anal. Chim. Acta* 101, 239 (1978).
106 M. S. Mohan and R. G. Bates, *Clin. Chem.* 21, 864 (1975).
107 M. S. Mohan and R. G. Bates, *NBS Spec. Publ. (U.S.)* 450, 293 (1977).
108 G. J. Moody, R. B. Oke and J. D. R. Thomas, *Analyst* 95, 910 (1970).
109 G. J. Moody and J. D. R. Thomas, *Chem. and Ind.* 1974, 644.
110 G. J. Moody and J. D. R. Thomas, *Lab. Pract.* 27, 285 (1978).

111 G. J. Moody and J. D. R. Thomas, Poly(vinyl chloride) matrix membrane ion-selective electrodes, in *Ion-Selective Electrode Methodology*, Vol. 1 (ed. A. K. Covington), CRC Press, Boca Raton (1979), p. 111.

112 G. J. Moody and J. D. R. Thomas, Poly(vinyl chloride) matrix membrane ion-selective electrodes, Chapter 4 of *Ion-Selective Electrodes in Analytical Chemistry* (ed. H. Freiser), Vol. 1, Plenum Press, New York (1978).

113 G. J. Moody and J. D. R. Thomas, Properties of ion-sensing membranes based on PVC matrices, in *Ion-Selective Electrodes* (ed. E. Pungor and I. Buzás), Symposium 1976, Akadémiai Kiadó, Budapest (1977), p. 41.

114 W. E. Morf, E. Lindner and W. Simon, *Anal. Chem.* 47, 1596 (1975).

115 S. D. Moss, J. Janata and C. C. Johnson, *Anal. Chem.* 47, 2238 (1975).

116 S. D. Moss, J. Janata and C. C. Johnson, *IEEE Trans. Biomed. Eng.* 25, 49 (1978).

117 K. Nagy, T. A. Fjeldly and J. S. Johannessen, Aspects of the LaF_3 ion-selective electrode, in *Ion-Selective Electrodes* (ed. E. Pungor and I. Buzás), Conference 1977, Akadémiai Kiadó, Budapest (1978), p. 491.

118 L. E. Negus and T. S. Light, *Instrum. Technol.* 19, 23 (1972).

119 U. Oesch and W. Simon, *Anal. Chem.* 52, 692 (1980).

120 Orion Research Newsletter, *Specific Ion Electrode Technology* 1, No. 5 (1969).

121 L. Pataki, Applicability of the ion-selective electrodes from the point of view of the thermodynamics, in *Ion-Selective Electrodes* (ed. E. Pungor and I. Buzás), Symposium 1976, Akadémiai Kiadó, Budapest (1977), p. 177.

122 E. Pungor, *Anal. Chem.* 39, 28A (1967).

123 E. Pungor and K. Tóth, *Anal. Chim. Acta* 47, 291 (1969).

124 E. Pungor and K. Tóth, *Analyst* 95, 625 (1970).

125 E. Pungor and K. Tóth, Ion-selective electrodes in non-aqueous solvents, in *The Chemistry of Nonaqueous Solvents* (ed. J. J. Lagowski), Vol. 5A, Academic Press, New York (1978).

126 E. Pungor and K. Tóth, *Pure Appl. Chem.* 36, 441 (1973).

127 E. Pungor and K. Tóth, Precipitate-based ion-selective electrodes, Chapter 2 of *Ion-Selective Electrodes in Analytical Chemistry* (ed. H. Fresier), Vol. I, Plenum Press, New York (1978).

128 E. Pungor, K. Tóth and P. Gábor-Klatsmányi, *Hung. Sci. Instr.* 49, 1 (1980).

129 G. Ali Qureshi and J. Lindquist, *Anal. Chim. Acta* 67, 243 (1973).

130 A. U. Ramsing, J. Růžička, J. Janata and M. Levy, *Anal. Chim. Acta* 118, 45 (1980).

131 R. Rangarajan and G. A. Rechnitz, *Anal. Chem.* 47, 324 (1975).

132 M. D. Reboiras, H. Pfister and H. Pauly, *Biophys. Chem.* 9, 37 (1978).

133 G. A. Rechnitz, *Chem. and Eng. News* 12.6.1967, p. 146.

134 G. A. Rechnitz and N. C. Kenny, *Anal. Lett.* 2, 395 (1969).

135 G. A. Rechnitz and M. R. Kresz, *Anal. Chem.* 38, 1786 (1966).

136 G. A. Rechnitz, M. R. Kresz and S. B. Zamochnik, *Anal. Chem.* 38, 973 (1966).

137 G. A. Rechnitz and Z. F. Lin, *Anal. Chem.* 40, 696 (1968).

138 Recommendations for publishing manuscripts on ion-selective electrodes (prepared for publication by G. G. Guilbault), Commission on Analytical Nomenclature, Analytical Chemistry Division, IUPAC, *Ion-Sel. El. Rev.* 1, 139 (1979).

139 R. A. Robinson and R. G. Bates, *Anal. Chem.* 45, 1684 (1973).

140 R. A. Robinson and R. G. Bates, *Marine Chem.* 7, 281 (1979).

141 R. A. Robinson, W. C. Duer and R. G. Bates, *Anal. Chem.* 43, 1862 (1971).

142 J. W. Ross, Solid state and liquid membrane ion-selective electrodes, Chapter 2 of *Ion-Selective Electrodes* (ed. R. A. Durst), NBS Special Publication No. 317, Washington (1969).

143 J. W. Ross, J. H. Riseman and J. A. Krueger, *Pure Appl. Chem.* **36**, 473 (1973).
144 J. Růžička and E. H. Hansen, *Anal. Chim. Acta* **63**, 115 (1973).
145 J. Růžička, E. H. Hansen and J. C. Tjell, *Anal. Chim. Acta* **67**, 155 (1973).
146 J. Růžička and C. G. Lamm, *Anal. Chim. Acta* **53**, 206 (1971).
147 J. Růžička, C. G. Lamm and J. C. Tjell, *Anal. Chim. Acta* **62**, 15 (1972).
148 J. Růžička and J. C. Tjell, *Anal, Chim. Acta* **51**, 1 (1970).
149 J. P. Sapio, J. F. Colaraotolo and J. M. Bobbitt, *Anal. Chim. Acta* **67**, 240 (1973).
149a J. Šenkýř and J. Petr, *Chem. Listy* **73**, 1097 (1979).
150 W. Severinghaus and A. F. Bradley, *J. Appl. Physiol.* **13**, 515 (1958).
151 A. Shatkay, *Anal. Chem.* **39**, 1056 (1967).
152 A. Shatkay, *Anal. Chem.* **48**, 1039 (1976).
153 B. Siramizu, J. Janata and S. D. Moss, *Anal. Chim. Acta* **108**, 151 (1979).
154 J. G. Schindler, *Biomed. Technik* **20**, 75 (1975).
155 J. G. Schindler, *Biomed. Technik* **24**, 203 (1979).
156 J. G. Schindler and W. Riemann, *Biomed. Technik* **20**, 75 (1975).
157 J. G. Schindler, G. Stork, H. J. Struh, W. Schmid and K. D. Karaschinski, *Fresenius Z. Anal. Chem.* **295**, 248 (1979).
158 R. W. Schmid and C. N. Reilley, *J. Am. Chem. Soc.* **78**, 5513 (1966).
159 J. Siemroth, I. Hennig and R. Claus, All solid state ion-selective electrodes prepared from fused sensing materials, in *Ion-Selective Electrodes* (ed. E. Pungor and I. Buzás), 2nd Symposium 1976, Akadémiai Kiadó, Budapest (1977), p. 185.
160 E. M. Skobets, L. I. Makovetskaya and Yu. P. Makovetskii, *Ikr. Khim. Zh.* **40**, 1132 (1974).
161 E. M. Skobets, L. I. Makovetskaya and Yu. P. Makovetskii, *Zh. Anal. Khim.* **29**, 845 (1974).
162 R. L. Smith, J. Janata and R. J. Huber, *J. Electrochem. Soc.* **127**, 1599 (1980).
163 *Theory, Design and Biomedical Applications of Solid State Chemical Sensors* (eds. P. W. Cheung, D. G. Fleming, M. R. Neuman and W. H. Ko), CRC Press, West Palm Beach (1978), p. 164.
164 R. C. Thomas, *Ion-Sensitive Intercellular Microelectrodes. How to Make and Use Them*, Academic Press, London (1978).
165 R. C. Thomas, *J. Physiol (London)* **220**, 55 (1972).
166 E. Ujec, O. Keller, N. Kříž, V. Pavlík and J. Machek, *Bioelectrochem. Bioenerg.* **7**, 363 (1980).
167 E. Ujec, O. Keller, N. Kříž, V. Pavlík and J. Machek, Double-barrel ion-selective K^+, Ca^{2+}, Cl^--coaxial microelectrodes (ISCM) for measurements of small and rapid changes in ion activities, in *Ion-Selective Microelectrodes and Their Use in Excitable Tissues* (ed. E. Syková, P. Hník and L. Vyklický), Plenum Press, New York (1981), p. 41.
168 E. Ujec, O. Keller, J. Machek and V. Pavlík, *Pflügers Archiv* **382**, 189 (1979).
169 U. K. Patent No. 1150698, April 30, 1969.
170 U.S. Patent No. 3442782, May 6, 1969.
171 R. E. Van de Leest, *Analyst* **101**, 433 (1976).
172 R. E. Van de Leest, *Analyst* **102**, 509 (1977).
173 R. E. Van de Leest and A. Geven, *J. Electroanal. Chem.* **90**, 97 (1978).
174 J. C. Van Loon, *Anal. Chim. Acta* **54**, 23 (1971).
175 J. C. Van Loon, *Int. J. Environ. Anal. Chem.* **3**, 53 (1973).
176 J. Veselý, O. J. Jensen and B. Nicolaisen, *Anal. Chim. Acta* **62**, 1 (1972).
177 J. Veselý and J. Jindra, Ionic electrodes on single crystal basis and with solid internal contacts, *Proc. IMEKO Symposium on Chemical Sensors*, Veszprém, Akadémiai Kiadó, Budapest (1968), p. 69.

178 Y. G. Vlasov, D. E. Hackleman and R. P. Buck, *Anal. Chem.* 51, 1570 (1979).
179 Yu. G. Vlasov, Yu. A. Tarantov and V. P. Letavin, *Zh. Prikl. Khim.* 53, 2345 (1980).
180 D. Vofsi and J. Jagur-Grodzinski, *Naturwiss.* 61, 25 (1974).
181 F. Vyskočil and N. Kříž, *Pflügers Archiv* 337, 265 (1972).
182 K. Vytřas, M. Dajková and M. Remeš, *Českosl. Farmacie* 2, 61 (1981).
183 J. L. Walker, *Anal. Chem.* 43, 89A (1971).
184 J. L. Walker, Single cell measurement with the ion-selective electrodes, Chapter 4 of *Medical and Biological Applications of Electrochemical Devices* (ed. J. Koryta), John Wiley and Sons, Chichester (1980).
185 J. L. Walker and H. M. Brown, *Physiol. Rev.* 57, 729 (1977).
186 J. H. Wang and E. Copeland, *Proc. Nat. Acad. Sci.* 79, 1909 (1973).
187 R. Wawro and G. A. Rechnitz, *Anal. Chem.* 46, 806 (1974).
188 J. N. Zemel, *Anal. Chem.* 47, 255A (1975).
189 J. N. Zemel, *Res./Dev.* 28, 38 (1977).

5

EXPERIMENTAL TECHNIQUES

Measurement with ion-selective electrodes appears at first sight extremely simple and this is one of the main reasons for the great popularity of the method. However, to obtain meaningful results a number of conditions must be met, conditions which are sometimes contradictory and difficult to fulfil. A severe limitation is imposed on measurements with ISEs because of a fact inherent to most electrochemical methods: namely, that the measurement depends on heterogeneous reactions occurring at the electrode–solution interface. Consequently, the reproducibility and long-term constancy of the conditions at the interface is of paramount importance for accurate and reproducible measurements. There is no general solution to this problem and an ideal state can be approached more or less closely only by judicious selection of the experimental conditions, starting with the sample preparation, through the actual measurement, to the handling of the results. This is the main reason for the fact that of the extensive literature on ISEs, only a small part deals with real practical applications rather than with the laboratory study of the electrodes and methods, and that routine use of the ISEs in chemical analysis is less common than it would appear from the literature.

To attain satisfactory performance with most electrochemical methods, including potentiometry with ISEs in routine analytical work, a certain amount of information on electrochemistry and experimental experience is required. In this respect, electrochemical methods often compare unfavourably with spectral methods. Nevertheless, in certain fields they are indispensable; the operator must be able to use them in such places and in such ways that they yield the required information more readily than other analytical methods.

5.1   Principal conditions for correct measurement with ISEs

5.1.1   *Sample preparation*

Measurement with ISEs requires that the test substances be present in solution. Because ionic species are being measured (except in the case of sensors employing auxiliary reactions, such as gas probes and enzyme electrodes), the solvent must be polar and water is usually used. Some determinations, mostly precipitation titrations, are performed in mixed media (water-alcohol, water-dioxan, etc.) to improve the determination limit.

To obtain a sample solution, gaseous samples are usually passed at a controlled flow-rate through a suitable absorption solution. For example, when determining the concentration of sulphur dioxide in the atmosphere, the air is passed through a tetrachloromercurate solution and sulphur dioxide is trapped as a result of the reaction

$$HgCl_4^{2-} + 2SO_2 + 2H_2O \rightleftharpoons Hg(SO_3)_2^{2-} + 4Cl^- + 4H^+ \qquad (5.1)$$

The absorption solution contains EDTA to stabilize the $HgCl_4^{2-}$ complex ion, and glycerol to suppress the oxidation of SO_2; at pH 6.9 the absorption is virtually complete [74]. For determination of SO_2 at very low concentrations, the air is passed through a filter impregnated with this absorption solution [8]; the sampling time can then be longer and the gas flow-rate higher than with the absorption solution alone. Another example is the monitoring of atmospheric pollution with fluoride using an absorption solution containing 0.1M NaOH [79].

Sometimes substances to be determined in gaseous samples are adsorbed on solid adsorbents and then transferred to solution. Thus nitrogen oxides present in the atmosphere are chemisorbed on lead dioxide heated to a temperature of 190 °C. The lead nitrate formed is extracted into hot water and the NO_3^- anion is then determined [29].

When substances adsorbed on aerosol particles are to be determined, the gas is passed through a membrane or other filter and the filter is dissolved in or extracted with a suitable solution. An interesting method is used for determination of fluoride adsorbed on atmospheric aerosols [87]. The particles are trapped on a filter impregnated with citric acid and heated to 80 °C, while the fluorides pass through and are absorbed in a thin layer of sodium carbonate in a spiral absorber. The sodium carbonate is periodically washed with a sodium citrate solution, in which solution the fluoride is then determined, and the absorption layer regenerated.

If solid samples are insoluble in water, some decomposition procedure must be used. For inorganic materials, decomposition with mineral acids is most often employed (for a survey of decomposition techniques see [33]). When the sample cannot be dissolved in an acid, it can either be fused (most often with alkali carbonates, hydroxides or their mixtures [157, 47]) or sintered (usually with mixtures of alkali carbonates with divalent metal oxides, sometimes in the presence of oxidants [54]). Sintering is usually preferable, because then contamination of the sample and the resultant ionic strength are lower than is the case with fusion. Solid–liquid extraction, distillation and pyrohydrolysis [161] can be useful, because they are selective and can simultaneously lead to pre-concentration of the test substance.

Organic samples are usually mineralized to render them soluble in water and to eliminate possible interference from the matrix (for example through adsorption on the electrode surface). Mineralization can be carried out by heating with strong oxidizing acids, by fusion with alkalis or by combustion in an oxygen atmosphere [142]. Blood samples must be treated with heparin if they are not immediately analysed and the content of O_2 and CO_2 must be carefully maintained; otherwise the dissociation equilibrium of calcium is shifted [78, 96].

Samples of natural water should either be analysed immediately or be stored (not for a very long time) at a decreased temperature to suppress microbial processes. For the determination of nitrate and nitrite it is useful to conserve the samples by addition of 1 ml chloroform or 0.1% phenylmercuric acetate per litre. To prevent oxidation of sulphide and some other substances in water samples, reductants are added [5, 147]. If the distribution of a species between the free ionic form and various complexes is to be studied, as is often the case, care must be taken not to shift the equilibrium by adding substances that would enter into side reactions with the studied species.

5.1.2   Adjustment of the experimental conditions

The sample solution must meet certain requirements for precise and accurate measurement. The most important parameters affecting electrode performance are the pH, the ionic strength, the presence of interferents and the temperature.

The solution pH affects the function of all ISEs, either through interference of hydroxonium or hydroxide ions in the membrane reaction (for example the interference of OH^- with the function of fluoride ISEs at pH values greater than about 5.5), or through chemical interference in solution (for example, formation of poorly dissociated HF and HF_2^- in acid solutions, which are not sensed by a fluoride ISE; or formation of HS^- and H_2S with decreasing pH, which are not sensed by a sulphide ISE), or both. Moreover, the pH value can affect the equilibria of interferents in the solution. The pH must thus be adjusted with all these effects in mind. Fortunately, it is usually sufficient to maintain the pH within a certain region rather than at a single precise value.

The ionic strength of the solution influences the activity coefficients of the species present and the values of the liquid–junction potentials in the system and thus must be held constant in all related measurements, within a range of about 0.1 to 2M.

If the interferents are present at concentrations so high that they cause significant determination error, they must either be removed or masked. Removal of interfering substances can sometimes be achieved during the preparation of the sample solution (see section 5.1.1), for example by selective absorption of the gaseous test substance in the solution, using a selective fusion or sintering procedure, etc.; or a separation step (ion-exchange, extraction etc.) can be included. However, separation procedures always make the procedure more complicated and time consuming, thus offsetting one of the main practical advantages of potentiometry with ISEs, namely, its simplicity and rapidity, and introduce additional errors in the determination through contamination of the sample by the chemicals used and loss of the determinand due to adsorption on the vessel walls, hydrolysis, volatilization, redox reactions, etc. On the other hand, some separation processes can simultaneously be used for pre-concentration of the test substance.

Masking of interferents in the test solution by suitable chemical reactions is usually preferable. It is advantageous to prepare a solution of a masking agent in such a way that it can simultaneously adjust the optimal pH and ionic strength for the potentiometric measurement. A classical example of such an agent is TISAB (total ionic strength adjustment buffer) [41] used in the determination of fluoride, containing an acetate pH-buffer, nitrate to adjust the ionic strength, and citrate to mask cations, especially Fe^{3+} and Al^{3+}, which interfere by formation of fluoride complexes. Masking agents often adversely affect electrode function, especially the response rate and sensitivity, by interacting with the electrode membrane. For this reason, the original TISAB composition has been modified by various researchers, for example by replacing the citrate by DCTA [53]. Another example of such a multifunctional agent is CAB (complexing antioxidant buffer) [147] used in the determination of copper in water. It contains an acetate pH-buffer, a fluoride solution and formaldehyde, and maintains a constant ionic strength, complexes interfering metals (mainly Fe^{3+}) and has a mild reducing effect. There are many similar multifunctional agents (for a more detailed description see [158]).

The problem of selectivity can often be efficiently solved when using flow techniques, such as continuous-flow analysis or flow–injection analysis (see section 5.3).

If the measuring error caused by temperature fluctuations is not to exceed an acceptable level, the temperature should be held constant within about ±0.5 °C; in measurements where a high precision is required (for example the determination of equilibrium constants), the solutions must be thermostatted at least within ±0.1 °C. It is recommended [158] that the measurement generally be carried out at a somewhat elevated temperature (at 25 to 30 °C), as the electrode response is faster and precision of measurement is improved.

It is further useful to measure ionic species in stirred or flowing solutions, because the electrode response is then faster, the determination limit is often better than in quiescent solutions and the measurement precision is also improved These improvements apparently result from the effect of solution movement on film diffusion at the electrode surface, which is assumed to be the response-rate determining step [92, 154]. An obvious requirement is that the solution velocity and the cell geometry be constant.

Proper function of ISEs also requires pre-conditioning. The response of electrodes not preconditioned is usually slow and hard to reproduce. Pre-conditioning of ISEs usually involves soaking the electrode in a solution of the ion to be sensed at a concentration approximately in the middle of the attainable measuring range (i.e. about 10^{-3} M in most cases), followed by repeated measurement at various concentrations of the test ion, until the response is rapid and reproducible. Many electrodes exhibit a 'memory' effect, i.e. changes from more concentrated to less concentrated solutions yield temporary higher readings, and changes from less concentrated to more concentrated solutions temporary lower readings. This effect, apparently connected with test ion adsorption–desorption phenomena at the electrode surface, can be more or less suppressed by stirring the solution and by avoiding successive measurements of samples with very different test ion concentrations.

The membrane surface may become passivated by some solution components that are strongly adsorbed. This effect is often encountered in measurements on biological fluids containing proteins. These adsorption effects can sometimes be prevented by selecting a suitable composition of the sample and standard solution; for example by adding trypsin and triethanolamine to dissolve proteins [108]. Passive electrodes can sometimes be reactivated by soaking in suitable solutions (for example pepsin in 0.1 M HCl [68]) and in more serious cases the membrane must be replaced or a solid membrane be repolished.

5.1.3 Requirements of the measuring system

As electrode systems involving ISEs usually have a high resistance, up to several GΩ, the meter should have the highest possible input impedance. To obtain a measuring error of 0.1%, the input impedance of the meter should be 10^3 times the cell resistance. To keep the overall measuring error at a level of a few per cent, the meter should have a resolution of the order of 0.1 mV. Modern instruments readily comply with these requirements. For a useful view of these aspects see [21].

In measurements on high-impedance cells, noise is generated from electromagnetic fields and capacitance effects. Shielded cables that are as short as possible and not coiled or twisted are imperative. The instrument should not be placed close to electric engines, line voltage installations, etc., as the noise stemming from these sources cannot be effectively eliminated even by placing the apparatus in a Faraday cage. The resistance of the system against the ground must also be as high as possible.

An effective solution to these problems can be achieved by converting the high-impedance signal of the sensor to a low-impedance output using an electronic circuit placed directly in the ISE body [75]. In this respect, ion-selective sensors based on field-effect transistors are advantageous, as they provide a low-impedance signal without using additional circuits. For transmitting the signal greater distances, which might be required when using automated on-line industrial analysers or pollution monitors, digitization of the signal is useful to prevent the loss of the signal in the noise [75].

The precision and accuracy of the measurement also depend strongly on the reference electrode, which affects the results through fluctuations in its own potential and through the liquid–junction potential at the test solution–liquid bridge interface. This subject is extensively treated in [158]. Common electrodes of the second kind have sufficiently stable potentials at a constant temperature, but difficulties can be encountered due to temperature hysteresis. Silver chloride electrodes are preferable to calomel electrodes, because their temperature hysteresis is substantially smaller; with a calomel electrode, potential stabilization after a change in the temperature may even take several hours. Negligible temperature hysteresis is exhibited by the thallamide reference electrode [26, 158], which, however, has not yet found wider practical use.

When a constant ionic strength of the test solution is maintained and the reference electrode liquid bridge is filled with a solution of a salt whose cation and anion have similar mobilities (for example solutions of KCl, KNO_3 and NH_4NO_3), the liquid-junction potential is reasonably constant (cf. p. 24–5). However, problems may be encountered in measurements on suspensions (for example in blood or in soil extracts). The potential difference measured in the suspension may be very different from that obtained in the supernatant or in the filtrate. This phenomenon is called the suspension (Pallmann) effect [110] The appearance of the Pallmann effect depends on the position of the reference electrode, but not on that of ISE [65] (i.e. there is a difference between the potentials obtained with the reference electrode in the suspension and in the supernatant). This effect has not been satisfactorily explained; it may be caused by the formation of an anomalous liquid–junction or Donnan potential. It must be taken into consideration especially in biological and clinical measurements and it may sometimes be more suitable to avoid the use of a reference electrode by measuring a differential signal with two ISEs (see section 5.2.4).

5.2 Discontinuous potentiometric measurement
5.2.1 Determination of activities

ISEs respond to the activities of ions. To prepare activity standards, the individual activity coefficients of the pertinent ions must be known. However, individual activity coefficients cannot be determined accurately and can only be calculated approximately. For a discussion of conventional activity scales see p. 73–6.

When using standard solutions prepared on the basis of activities calculated from these activity scales, and provided there is no interference and that the prelogarithmic term in the E versus log a_X dependence is Nernstian or at least accurately known and constant, the sample activity can be determined from the ISE potentials obtained in the sample and in a standard solution (see (4.13), p. 74–6).

For accurate measurement, the liquid–junction potentials in the sample and in the standard must be identical or, at least, a correction must be made for the liquid–junction potential difference.

If measurements are to be carried out at low activities (for example in studying complexation equilibria), standard solutions cannot be prepared by simple dilution to the required value because the activities would irreproducibly vary as a result of adsorption effects, hydrolysis and other side reactions. Then it is useful to use well-defined complexation reactions to maintain the required metal activity value [14, 50, 132]. EDTA and related compounds are very well suited for this purpose, because they form stable 1:1 complexes with metal ions, whose dissociation can be controlled by varying the pH of the solution. Such systems are often termed metal-ion buffers [50] (cf. also p. 77) and permit adjustment of metal ion activities down to about 10^{-20} M. (Strictly speaking, these systems are defined in terms of the concentration, but from the point of view of the experimental precision, the difference between the concentration and activity at this level is unimportant.)

Finally it should be emphasized that to obtain unbiased activity values in samples, all operations affecting the activity coefficients (dilution, changes in the composition) must be avoided.

5.2.2 Determination of concentrations using calibration curves

Concentration values are more often required in analytical practice than activities (except for certain fields in clinical analysis and metal ion speciation in environmental control). Theoretically, the activity values obtained by measurement with ISEs could be recalculated to give concentrations, using the activity coefficients calculated as described in section 5.2.1. However, this would not only be tedious and imprecise but in most cases practically impossible, because the sample solution composition is complex, owing to the complicated matrices of samples and to the necessity of adding various reagents to improve sample stability and the measuring selectivity (see section 5.1).

If the sample solutions contain an indifferent electrolyte at a sufficiently high and constant concentration, then it can be assumed that the activity coefficients and liquid–junction potentials are also constant and can be included in the value of the 'standard potential' of the given ISE. It is then possible to calibrate the system using solutions with known concentrations of the ion sensed by the ISE and thus to obtain sample concentration values directly. A general condition is that the compositions of the sample and standard solutions be as close as possible, (especially the pH, the ionic strength and possible interferents).

The simplest but also the least reliable calibration method is the use of a single standard solution. The electrode response is assumed to be Nernstian. The slope of the potential versus concentration dependence can also be determined experimentally, by using two standard solutions with different concentrations. To avoid large errors, the standard concentration should be as close as possible to the sample concentration in calibration with a single standard solution.

A much more reliable method involves the use of two standards. From the point of view of the measuring precision, it is best when the potential measured in the sample solution, E_x, lies midway between the potentials obtained in the two standard solutions, E_1 and E_2. Consequently, for the corresponding concentrations it holds that $c_x = (c_1 c_2)^{1/2}$. Provided that all samples measured fall in the interval between the two standards and that the slope of the potential versus concentration dependence is constant within this interval, this method usually gives results satisfactory for routine analytical determinations.

When the test component content in the samples varies over a wider interval, a calibration curve must be constructed. Calibration curves with ISEs are usually linear over several concentration orders (usually from about 10^{-2} M to about 10^{-5} M) and their slope is close to the theoretical Nernstian value. Both at high and at low concentrations with respect to the linear part, the calibration dependence becomes curved and, eventually, independent of the test substance concentration (see fig. 5.1). The upper limit of the ISE response is mostly given by saturation of the active sites in the electrode membrane (for example ion-exchange sites), whereas the lower limit is determined by solubility of the

membrane, by test substance present in the chemicals, by adsorption of the
test substance or of other species on the membrane, etc. When an interfering
compound is present in the solution, the linear portion of the curve can be
considerably narrowed (see fig. 5.1, curves 2 and 3).

It is desirable that the test substance concentrations in the samples lie
within the linear part of the calibration curve, in order to keep the standard
deviation of the measurement at an acceptable level. More concentrated sample
solutions can be appropriately diluted, while a determinate amount of the test
component can be added to less concentrated solutions to bring the total concentration up to the linear part of the calibration curve. However, the latter
approach is of only limited importance because the presence of a constant
amount of the test component may cause the relative changes in the sample
concentration to become too small to measure precisely; therefore, it is useful
only when calibration curve slope varies rapidly with changing concentration [144].

On the other hand, it is possible to measure even in non-linear regions of the
calibration curve; if the optimal measuring conditions are carefully maintained,
the relative error of measurement usually does not exceed about 20%. At very low
concentrations, semiquantitative procedures can be employed; for example, the
sample is compared with standards and the direction of the drift of the unstabi-
lized potential indicates whether the sample concentration is higher or lower
than that in the standard [147, 162]. It is necessary to bear in mind that the
ISE response at very low concentrations is generally slow and the potential is
unstable, so that potential values read after a certain, fixed time interval must
often be used instead of stabilized values.

Fig. 5.1. Typical ISE calibration curves: 1 – curve in the absence of
interferents; 2,3 – curves in the presence of an interferent; curve 3
corresponds to a higher interferent concentration.

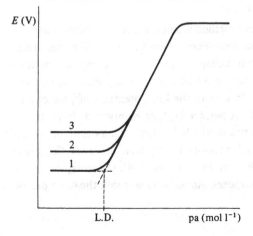

There have also been attempts to calculate the theoretical ISE response at low concentrations, where the calibration curve is non-linear, assuming that the ISE response is limited by the membrane dissolution [93, 151] (cf. section 3.2). For an isovalent membrane (i.e. a membrane with identical charge numbers of ion-exchanging site and determinand) the equation (cf. (3.4.13))

$$E_{ISE} = \text{const} + S \log \{\tfrac{1}{2}[c_{J^+}^0(1)] + [\tfrac{1}{4}c_{J^+}^0(1)^2 + P_{JA}]^{1/2}\} \qquad (5.2)$$

was derived (cf. [93]), where S is the Nernstian slope, $c_{J^+}^0(1)$ is the test ion concentration and P_{JA} is the membrane solubility product. A similar equation was derived [93] for calibration curve non-linearity caused by interference arising when a complexing agent reacts with one ion of a dissolving crystalline membrane, whereas the ISE responds to the other ion in the membrane:

$$E_{ISE} = \text{const} + S \log [\tfrac{1}{2}c_{J^+}^0(1) + \tfrac{1}{4}c_{J^+}^0(1)^2 + K']^{1/2}, \qquad (5.3)$$

where

$$K' = P_{JA}(1 + \beta[L]) \qquad (5.4)$$

and $c_{J^+}^0(1)$ is the concentration of the ion of the membrane sensed by the ISE, P_{JA} is the membrane solubility product, β is the overall stability constant of the complex formed and [L] is the ligand concentration.

These calculations are based on rather simplified assumptions and thus the values obtained may frequently differ considerably from experimental values.

The limit of determination is commonly estimated by finding the intercept of extrapolated linear parts of the calibration curve (see point L.D. in fig. 5.1). However, it is often difficult to construct a straight line through the experimental potentials at low concentrations and, moreover, the precision of the potential measurement cannot be taken into consideration. Therefore, it has been recommended that, by analogy with other analytical methods, the determination limit be found statistically, as the value differing with a certain probability from the background [94].

In general, the calibration curve method is suitable for all samples where the test substance is not bound in complexes or when it can be liberated from complexes by suitable sample pretreatment. Otherwise, the compositions of the samples and of the standard solutions must be as similar as possible to obtain results with acceptable accuracy. In view of the ISE potential drift, the calibration must be repeated often (at least twice a day). As mentioned above, the precision of the determination is not particularly high; with a common precision of the potential measurement at a laboratory temperature of ±1 mV the relative error is ±4% for univalent and ±8% for divalent ions [58]. However, this often suffices for practical analytical purposes. An advantage is that the same precision is obtained over all the linear part of the calibration curve, owing to the logarithmic relationship between the ISE potential and the test substance activity (concentratioi

5.2.3 *Addition techniques and potentiometric titration*

In addition techniques, the test substance concentration is determined from the difference in the ISE potentials obtained before and after a change in the sample solution concentration. The main advantage lies in the fact that the whole measurement is carried out in the presence of the sample matrix, so that results with satisfactory accuracy and precision can be obtained even if a substantial portion of the test substance is complexed. Several addition techniques can be used, namely, single, double or multiple known addition methods, in which the sample concentration is increased by additions of a test substance standard solution; single, double or multiple known subtraction methods, in which the sample concentration is decreased by additions of a standard solution of a substance that reacts stoichiometrically with the determinand; and analyte addition and subtraction methods, in which the sample is added to a test substance solution or to a reagent solution.

To obtain reliable results with addition techniques, several conditions must be satisfied:

(*a*) The activity coefficient of the test ion and the liquid–junction potential must not change during the addition; this condition is met by maintaining constant ionic strength.

(*b*) The fraction of the complexed test substance must be constant; there-fore, the complexing agent concentration must be sufficiently high and constant, and a constant solution pH must be maintained. When the concentrations of the test substance and of a complexing agent are comparable in the sample, the complexation equilibrium can easily be disturbed. It is then advisable either to completely release the substance from the complex by a suitable chemical reaction, or to add a sufficient excess of a suitable complexing agent.

(*c*) The slope of the potential versus concentration dependence is constant (preferably Nernstian).

(*d*) There are no interferences.

(*e*) The change in the solution volume during addition is negligible.

The simplest addition methods are the known addition method (KAM) and known subtraction method (KSM). Two potential values are obtained in the sample solution, before the addition (E_1) and after the addition (E_2) of volume V_a of a standard solution to sample volume V_0. The sample concentration is c_x, the standard solution concentration is c_s. In KAM, the standard solution contains the determinand. In KSM, the standard solution contains substance B that reacts with determinand A according to the equation,

$$aA + bB \rightarrow pP, \tag{5.5}$$

where B and P are not sensed by the ISE. The reaction ratio between A and B is $a/b = \mu$.

Provided that the determinand activity coefficient and the fraction that is complexed remain constant during measurement, then

$$E_1 = E_0 \pm \left(\frac{RT}{zF}\right) \ln c_x \tag{5.6}$$

and

$$E_2 = E_0 \pm \left(\frac{RT}{zF}\right) \ln \left(\frac{c_x V_0 + c_s V_a}{V_0 + V_a}\right) \quad \text{(KAM)} \tag{5.7}$$

or

$$E_2 = E_0 \pm \left(\frac{RT}{zF}\right) \ln \left(\frac{c_x V_0 - \mu c_s V_a}{V_0 + V_a}\right) \quad \text{(KSM).} \tag{5.8}$$

Subtracting (5.6) from (5.7) or (5.8) and substituting the value, $S = \pm(RT/zF)$ ln 10, yields

$$\Delta E = E_2 - E_1 = S \log \left(\frac{c_x V_0 + c_s V_a}{c_x(V_0 + V_a)}\right) \quad \text{(KAM)} \tag{5.9}$$

or

$$-\Delta E = E_2 - E_1 = S \log \left(\frac{c_x V_0 - \mu c_s V_a}{c_x(V_0 + V_a)}\right) \quad \text{(KSM).} \tag{5.10}$$

On taking antilogarithms and rearranging,

$$c_x = \frac{c_s V_a}{V_0 + V_a} \frac{1}{10^{\Delta E/S} - (V_0/V_0 + V_a)} \quad \text{(KAM),} \tag{5.11}$$

or

$$c_x = -\frac{\mu c_s V_a}{V_0 + V_a} \frac{1}{10^{-\Delta E/S} - (V_0/V_0 + V_a)} \quad \text{(KSM).} \tag{5.12}$$

Assuming that $V_0 \gg V_a$, for KAM,

$$c_x = \frac{\Delta c}{10^{\Delta E/S} - 1} \tag{5.13}$$

where

$$\Delta c = \frac{c_s V_a}{V_0}$$

and for KSM,

$$c_x = -\frac{\Delta c'}{10^{-\Delta E/S} - 1} = \frac{\Delta c' 10^{\Delta E/S}}{10^{\Delta E/S} - 1} \tag{5.14}$$

where

$$c' = \frac{\mu c_s V_a}{V_0}.$$

If slope S is not Nernstian, it must be replaced by an experimentally determined value, which can be obtained by measuring the E versus $\log c_x$ dependence at a constant temperature. A simpler method [105] involves the measurement of the ISE potential before (E_1) and after (E_2) dilution of the test solution with the blank solution. The experimental slope is then given by the equation

$$S_{ex} = \frac{\Delta E}{\log (V_2/V_1)}, \tag{5.15}$$

where ΔE is the potential difference $E_1 - E_2$, V_1 the original solution volume and V_2 the solution volume after dilution.

The result of the determination can then be calculated from (5.13) and (5.14), or from tables listing the $c_x/\Delta c$ and $c_x/\Delta c'$ ratios for various values of ΔE and S [101, 103, 104]. Finally, nomographs have been proposed for this purpose [67]. However, because the nomographs are insufficiently precise, numerical calculation, readily and rapidly performed with modern calculators and computers, is preferable.

The known addition method has an advantage in greater rapidity and somewhat improved precision compared with calibration with two standards. It is claimed that it requires about half the time necessary for two-standard calibration [17] and that the relative error is about ±2% [22]. A disadvantage is the necessity of determining the S_{ex} value for most systems. The method should be used only for concentrations corresponding to the linear part of the calibration curve. To eliminate the effect of a change in the ionic strength, it has been recommended [19] that a mixture of a standard with the test solution be added to the sample. The known subtraction method is much less used in practice.

The analyte addition method (AAM) involves adding the sample solution to a standard solution of the determinand, whereas in the analyte subtraction method (ASM) the sample is added to a standard solution of an ion that reacts stoichiometrically with the test substance and is sensed by an ISE. These methods are advantageous for determinations on small samples for which microelectrodes would otherwise have to be used. pH adjustment and masking of interferents in the sample is unnecessary because all these operations can be done beforehand on the standard solution. Furthermore, the analyte subtraction method widens the application of ISEs because ions for which an ISE is not available and various non-ionic substances can also be determined, provided they undergo a stoichiometric reaction with the ion present in the standard solution. For example, residual chlorine (a mixture of Cl_2, HOCl, OCl^- and organic chlorocompounds) in potable waters can readily be determined by ASM by adding the sample to a standard solution of iodide and monitoring the decrease in the iodide activity using an ISE [106] (see p. 136). These methods are also suitable for samples in which changes may occur on dilution.

The methods are mathematically analogous to KAM and KSM. If the volume of the standard solution is V_0, the volume of the sample added is $V_a(V_0 \gg V_a)$, the standard concentration is c_s, that of the sample c_x and the reaction ratio between the standard sensed and the sample added is μ, then

$$E_1 = E_0 \pm \left(\frac{RT}{zF}\right) \ln c_s \tag{5.16}$$

and

$$E_2 = E_0 \pm \left(\frac{RT}{zF}\right) \ln \left(\frac{c_s V_0 + c_x V_a}{V_0 + V_a}\right) \quad \text{(AAM)} \tag{5.17}$$

or

$$E_2 = E_0 \pm \left(\frac{RT}{zF}\right) \ln \left(\frac{c_s V_0 - \mu c_x V_a}{V_0 + V_a}\right) \quad \text{(ASM)}. \tag{5.18}$$

Following the same procedure as with KAM and KSM, the final equations are obtained in the form,

$$c_x = (c_s V_0 / V_a)(10^{\Delta E/S} - 1) = \Delta c (10^{\Delta E/S} - 1) \quad \text{(AAM)} \tag{5.19}$$

and

$$c_x = (c_s V_0 / \mu V_a)(1 - 10^{-\Delta E/S}) = \Delta c'(1 - 10^{-\Delta E/S}) \quad \text{(ASM)}. \tag{5.20}$$

The double and multiple addition methods are introduced in an attempt further to improve the measuring precision, because with three or more experimental potential values the slope value S need not be known. Under the same assumptions and with the same symbols as above, provided the same volumes are always added, it holds for the nth addition of the determinand standard solution that

$$E_{n+1} = E_0 \pm \frac{RT}{zF} \ln \frac{c_x V_0 + n c_s V_a}{V_0 + n V_a} \tag{5.21}$$

It can be seen that ratio R, given by

$$R = \frac{\Delta E_n}{\Delta E_{n-1}} = \frac{E_n - E_1}{E_{n-1} - E_1}$$

$$= \frac{\log \{[c_x V_0 + (n-1)c_s V_a]/c_x [V_0 + (n-1)V_a]\}}{\log \{[c_x V_0 + (n-2)c_s V_a]/c_x [V_0 + (n-2)V_a]\}} \qquad (5.22)$$

is the same for ions of different valences because the slope value has been eliminated. The c_x value can again be found in tables [105] for double known addition, assuming that $V_0 \gg V_a$. Generally, the calculation of c_x for multiple addition methods is tedious and computing techniques must be used. The computing procedures (see e.g. [4, 16, 42, 56, 85, 146, 167]) are mostly based on non-linear least-squares curve fitting and the value of V_a is not neglected. The $c_x V_0$ value should be greater than $c_s V_a$, but even under this condition the convergence of the procedure is sometimes slow. In optimal cases the values of E_0, S and c_x converge within three of four cycles. Some of these procedures (for example [4]) involve optimization of the volumes added during the measuring procedure.

Double addition methods do not significantly improve the precision of the determination and thus multiple addition methods should be preferred. A modification of double known addition is the known-addition known-dilution method, in which first a standard solution is added and then the solution is diluted, preferably to regain the initial concentration of the determinand [105].

Potentiometric titration is actually a form of the multiple known subtraction method. The main advantage of titration procedures, similar to multiple addition techniques in general, is the improved precision, especially at high determinand concentrations. ISEs are suitable for end-point indication in all combination titrations (acid–base, precipitation, complexometric), provided that either the titrand or the titrant is sensed by an ISE. If both the titrant and the titrand are electro-inactive, an electrometric indicator must be added (for example Fe^{3+} ion can be titrated with EDTA using the fluoride ISE when a small amount of fluoride is added to the sample solution [126]).

To obtain satisfactory results, several conditions must be fulfilled. The chemical reaction involved must be stoichiometric, no side reactions may occur or at least should be reproducible and quantitatively describable (such as the reaction of protons with ligands in chelatometric titrations) and the main reaction should have a high equilibrium constant value. The ISE should respond only to the titration system and the response should be fast and reproducible.

A theoretical treatment of combination titrations with an ideal indicator electrode was given by Meites *et al.* [89–91]. They have shown that the dilution effect causes a deviation of the titration curve inflection point from the equivalence point. However, this deviation is small compared with the error caused by interferences in the electrode response and by a low value of the equilibrium constant of the titration reaction, as has been shown [23, 24, 133, 134] for precipitation and complexometric titrations with ISEs.

If a titration is based on the combination reaction,

$$M^{n+} + X^{n-} \rightleftarrows MX, \tag{5.23}$$

where MX is either a precipitate with the solubility product

$$K_s = [M^{n+}] [X^{n-}] \tag{5.24}$$

or a complex with the dissociation constant,

$$K_d = [M^{n+}] [X^{n-}] / [MX] \tag{5.25}$$

and if M^{n+} is monitored, then the concentration of M^{n+} during the titration is given by [90, 91, 133, 134]

$$[M^{n+}] = \tfrac{1}{2} [\phi + (\phi^2 + 4K_s)^{1/2}] \tag{5.26}$$

for precipitation titration and

$$[M^{n+}] = \tfrac{1}{2} \{\chi + [\chi^2 + 4K_d c_M^0 / (1 + rx)]^{1/2}\} \tag{5.27}$$

for complexometric titration. Here c_M^0 is the sample initial concentration, x is the degree of titration defined as

$$x = c_x V_x / c_M^0 V_M^0, \tag{5.28}$$

where V_M^0 is the initial sample volume, c_x is the titrant concentration, V_x is the titrant volume required to attain the equivalence point and r is the dilution factor, given by,

$$r = c_M^0 / c_x. \tag{5.29}$$

Quantities ϕ and χ are given by the expressions,

$$\phi = (1 - x)c_M^0 / (1 + rx) \tag{5.30}$$

and

$$\chi = [(1 - x)c_M^0 / (1 + rx)] - K_d. \tag{5.31}$$

With an ideal electrode (5.26) and (5.27) could be used directly to compute the titration curve. However, with an ISE it is necessary to consider interferences and to calculate the potential from the Nikolsky equation. If it is assumed that interferents are present both in the sample and in the titrant, then the electrode potential is given by [133, 134],

$$E = E^{0\prime} + \frac{RT}{zF} \ln \left\{ [M^{n+}] + \sum_i \frac{k_i^{pot} c_i^0}{(1 + rx)^{z/n_i}} + \sum_i k_i^{pot} c_i^{0\prime} \left(\frac{rx}{(1 + rx)} \right)^{z/n_i} \right\} \tag{5.32}$$

where c_i^0 and $c_i^{0\prime}$ are the initial concentrations of the interfering ions in the sample and the titrant, respectively, and the n_i are their charges.

When the second derivative of (5.32) is calculated and set equal to zero, the inflection point of the titration curve is obtained [23, 24, 133, 134]. It has been found that the theoretical titration error generally increases with decreasing sample concentration, with increasing value of the solubility product or of the dissociation constant, with increasing value of the dilution factor and with increasing concentration of the interferents. Larger errors are obtained with unsymmetrical titration reactions. The overall error is a combination of these factors; the greatest effect is exerted by the sample concentration, a smaller one by the equilibrium constant and the interferents, and the smallest by dilution. To obtain errors below 1%, it must approximately hold that $c_M^0 \geqslant 10^{-2}$ M, $K \leqslant 10^{-8}$, $\sum_i k_i^{pot} c_i^0 \leqslant 10^{-3}$ to 10^{-4} and $r \leqslant 0.3$.

The titration error can often be decreased by titrating in non-aqueous or mixed media, in which the titration reaction products are more stable and the dissolution of the ISE membrane is sometimes suppressed, thus decreasing the limit of determination. Unfortunately, the electrode response in such media is usually slower [55].

An interesting indirect end-point detection method is based on the work of Siggia *et al.* [143], and Reilley and Schmid [119], who originally developed the technique for chelometric titrations of electro-inactive metals with a mercury indicator electrode. Assume that metal ion M is titrated by titrant Y (ionic charges are omitted for the sake of simplicity),

$$M + Y \to MY, \tag{5.33}$$

$$K_{MY} = [MY]/[M][Y]. \tag{5.34}$$

The indicator electrode responds to another metal ion, M′:

$$E = E_{M'}^{0'} + (RT/zF) \ln [M']. \tag{5.35}$$

Ion M′ also forms chelate M′Y with titrant Y, that is more stable than MY,

$$M' + Y \to M'Y \tag{5.36}$$

$$K_{M'Y} = [M'Y]/[M'][Y] \tag{5.37}$$

$$K_{M'Y} > K_{MY}. \tag{5.38}$$

If a small amount of chelate M′Y is added to the sample solution, the electrode potential during the titration is

$$E = E_{M'}^{0'} + \frac{RT}{zF} \ln \frac{[M][M'Y]K_{MY}}{[MY]K_{M'Y}} \tag{5.39}$$

and thus the electrode responds to variations in [M] and [MY]. A small amount of M′Y (10^{-4} to 10^{-5} M) suffices and need not be accurately measured. The cupric ion ISE is very well suited for this type of titration, because Cu^{2+} forms a very stable complex with EDTA (see [11, 122]).

A general disadvantage of potentiometric titration curves is their sigmoid shape, because the end-point must be located in the region of the greatest slope of the curve, where precision is poorest. It has been shown [2] that the worst results are obtained when locating the end-point on the unmodified curve, irrespective of which graphical method was used. Better results are obtained when plotting $\Delta E/\Delta V$ versus V. However, the most important points are still located around the inflection point and are subject to a large error. Therefore, Gran [45] constructed the $\Delta V/\Delta E$ versus V dependence. In this way, two approximately linear dependences are obtained, whose intercept is located on the volume axis and denotes the end-point. The points close to the end-point can then be neglected and the precision and accuracy are improved, especially when the straight lines are constructed using the linear regression method [2].

In potentiometry with ISEs, however, the second Gran method [46] has found especially wide use, not only in titrations, but also in multiple addition methods in general. In these methods, the concentration of the test substance is plotted against the volume of the titrant or of the standard solution and thus the curve is linearized. The end-point in the titration or the determinand concentration in a multiple addition method is found as the intercept of the straight line with the volume axis. Linearization is attained by taking the antilogarithm of the Nernst equation:

$$E = E^{0'} + S \log c, \tag{5.40}$$
$$10^{E/S} = 10^{E^{0'}/S} c = kc. \tag{5.41}$$

Hence, the Gran plot is obtained by plotting $10^{E/S}$ against the volume (see fig. 5.2). To facilitate this procedure, a special semi-antilogarithmic paper is

Fig. 5.2. Gran linearization of a potentiometric titration curve.

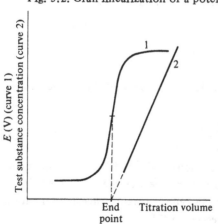

marketed by various manufacturers. For example, Orion Research Inc. [107] manufacture graph paper for a slope of 58 mV with a correction for 100% or 10% dilution, provided by skewing the coordinates upwards from the left to the right. A similar function is performed by the Gran slide rule [146] or by an operational amplifier antilog circuit connected to the meter output [155].

In the Gran linearization of a multiple addition method, (5.21) is rearranged to give

$$[V_0 + nV_a]10^{En+1/S} = 10^{E0/S}[c_xV_0 + nc_sV_a],\qquad(5.42)$$

which is the equation of a straight line,

$$y = bV_a + a \qquad(5.43)$$

where

$$y = [V_0 + nV_a]10^{En+1/S} \qquad(5.44)$$
$$b = 10^{E0/S}nc \qquad(5.45)$$

and

$$a = 10^{E0/S}c_xV_0. \qquad(5.46)$$

Hence, in the intercept with the horizontal axis (i.e. for $y = 0$) it holds that (see fig. 5.3)

$$nc_sV_x = -c_xV_0 \qquad(5.47)$$

and thus

$$c_x = -nc_s\frac{V_x}{V_0} \qquad(5.48)$$

Fig. 5.3. The use of the Gran method in the multiple addition method. For explanation see the text.

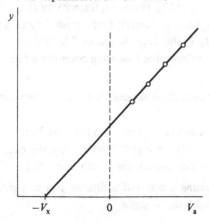

where V_x is the volume of the standard solution added that corresponds to the test concentration c_x.

A drawback of Gran plots is the fact that all deviations from the theoretical slope value cause an error and that side reactions are not considered. The method was modified by Ingman and Still [63], who considered side reactions to a certain degree, but the equilibrium constants and the concentrations of the components involved must be known. The Gran method is, however, advantageous for determinations in the vicinity of the determination limit. The extrapolation of the linear dependence yields the sum $c_x + c_r$, where c_r is the residual concentration of the test component produced by impurities, dissolution of the ISE membrane, etc.

An interesting method is single-point titration, in which a certain amount of reagent is added to the sample and the ISE potential is measured in the resultant solution [6, 7, 28, 62, 64, 66, 77]. The authors claim that the precision is similar to that of a classical titration (a relative standard deviation of 0.1 to 1.3%). Time is saved and the method is readily applicable to automated measuring systems.

The precision of addition techniques has been studied in many papers. It has been shown [88] that the lower the ratio $c_x V_0/c_s V_a$, the smaller the error. The effect of the slope uncertainty is unimportant at values $c_x V_0/c_s V_a < 1$, but its importance rapidly increases when the ratio increases above unity. Evaluation of the precision of these methods using computer simulation [57-60] has shown that the error depends on whether the slope is known or not and that in some cases the precision of multiple addition is surprisingly poor, worse than with a calibration curve or a single addition. The highest precision is generally obtained when using numerical curve fitting and multiparameter refinement [2]. On the other hand, in the numerical treatment a possible non-linearity of the $y = f(V)$ dependence, resulting from side reactions of the test substance or slow response of the ISE, can escape attention [18, 20, 112]. However, linearity of the $y = f(V)$ dependence need not mean that a systematic error, caused by changes in the ionic strength or by uncertainty in the slope, is absent. The highest precision and accuracy can be attained under the following conditions [18, 20, 112]:

(*a*) The standard concentration is as close as possible to the sample concentration.
(*b*) The electrode slope is known with good precision (at least 2%).
(*c*) When $1 < [V_{a,\max}(c_s + c_r)]/[V_0(c_x + c_r)] < 3$, where c_r is the background concentration (impurities, membrane dissolution, etc.).

The precision increases with increasing number of additions. For very precise determinations, at least 10 additions should be made.

5.2.4 Various modified techniques and automation of discontinuous potentiometric measurements

A significant increase in the measuring precision and accuracy can sometimes be attained by differential measurement with two ISEs. Two principal approaches can be taken:

(a) A cell without liquid junction is used, containing two different ISEs, one selective for the determinand and the other for another ion whose activity is maintained constant. The latter electrode thus has a constant potential and functions as a reference electrode. A glass electrode can be used for the purpose in acid-base buffered solutions or the fluoride ISE in solutions with a constant F^- activity [84].

(b) A cell with liquid junction is employed, containing two identical ISEs selective for the determinand; one ISE is immersed in a sample solution with concentration c_x and the other in a standard solution with concentration c_s. Assuming that the liquid-junction potential is constant, the potential difference measured is

$$\Delta E = \Delta E_0 + S_1 \log c_x - S_2 \log c_s. \qquad (5.49)$$

A technique utilizing the advantages of both differential measurement and titration is null-point potentiometry [37, 38, 83]. The same arrangement is used as described under (b), but concentration c_s is not constant and is varied until $c_s = c_x = c$. Then,

$$\Delta E = \Delta E_0 + (S_1 - S_2) \log c. \qquad (5.50)$$

With ideal electrodes, the ΔE should be zero, but with real electrodes this value must be determined by filling both half-cells with the same solution.

The differential technique described under (a) has an advantage in removal of the liquid-junction potential and of mechanical faults often encountered in work with reference electrodes of the second kind. The procedure described under (b) suppresses the potential fluctuations, but difficulties can arise from the very high resistance of a cell containing two ISEs. A differential amplifier was designed for this prupose [15]. The two ISEs used can also exhibit different slopes; electrode membranes were therefore prepared by cutting a single crystal into two halves, where each half contains a channel for passage of the solution and functions as an ISE [163].

The use of microelectrodes can be avoided in measurements on samples with microlitre volumes if the measurement is performed in a thin layer of the sample solution between a flat ISE and a plate-shaped reference electrode which are placed close together [153].

The great development in electronics and instrumentation has enabled automation of analytical processes. In discontinuous potentiometric measurements, the actual procedure is programmed and automated (see [4, 36, 87, 137–141, 146] for direct potentiometry and [43, 86] for titrations). Several electrodes can be used simultaneously: the potentials of up to five ISEs can be multiplexed, digitized and stored on a magnetic tape for subsequent computer handling [159, 160]. Microprocessor-controlled measuring instruments are common and are capable not only of storing the calibration data, calculating the results by the required method and displaying them in the required units, but can also correct the results for the background value. In handling the results, computing techniques are used extensively (see section 5.2.3 on the computer treatment of addition techniques). These procedures are often interactive [42, 43] and enable continuous optimization of the experimental conditions. The EFL (error function left) method has been developed [120, 121] for dealing with deviations of Gran dependences from linearity. The data set is divided into partially overlapping intervals of a suitable width. A regression straight line is constructed through each interval. These straight lines intersect the reagent (or standard) volume axis and the differences between the consecutive intercepts are plotted against the reagent volume. The minimum on this dependence corresponds to the theoretical linear part of the Gran transformation. The use of a new computer language, CONVERS, has been described [86] for end-point determination in potentiometric titrations and for fully automated generation of calibration curve data with multiple ISE systems.

In addition to automated analysers for general use, sophisticated single purpose instruments have been developed and marketed, chiefly for clinical analyses (for example Astra 4 and Astra 8 from Beckman for the determination of sodium and potassium in blood or Orion Space-Stat SS-20 or SS-30 for the determination of calcium in blood).

5.3 Measurements in flow systems

In analytical practice, measurements in flowing liquids are being used more and more extensively. There are three basic types of measurement, namely:

(*a*) Continuous monitoring, used chiefly in industrial analysers on production lines, in environmental control and in some cases in medicine (for example, following the level of certain species in blood during operations).

(*b*) Determinations in discrete samples aspirated into a flow system and segmented by air bubbles (continuous-flow analysis, CFA) or injected into a stream of a carrier medium (flow-injection analysis, FIA).

(*c*) Continuous detection in eluates from chromatographic columns.

Regardless of the analytical property measured and the measuring technique used, a continuous analyser or a detector should satisfy certain general requirements:

(a) The measuring sensitivity, defined as the slope of the calibration curve, should be high.

(b) The detection limit, defined statistically as a multiple of an estimate of the standard deviation of the noise (usually two to four times, depending on the significance level), should be low; in other words, the signal-to-noise ratio should be high.

(c) The signal should be stable and reproducible even in prolonged measurements.

(d) The linear dynamic range, i.e. the interval within which the calibration curve is linear, should be sufficiently wide.

(e) The response should be fast, i.e. the detection system should have a low time constant, so that the detector output signal provides an undistorted picture of the concentration input function. The total response time is the sum of the transport delay, i.e. the time required to transport the test substance from the sampling site to the active part of the detector, and the capacity delay, i.e. the time necessary for activation of the sensor. The requirement of a small time constant of the detector is especially stringent with chromatographic, CFA and FIA detectors and thus will be discussed in greater detail.

The separation efficiency in chromatography and the sample throughput in CFA and FIA are better the narrower the test substances zones. The experimental conditions must therefore be selected so that the zone broadening in the detection system is as low as possible. The zone width is conveniently expressed in terms of the time standard deviation, σ_t, or the volume standard deviation, σ_V, which are related by the expression,

$$\sigma_V = w\sigma_t, \tag{5.51}$$

where w is the volume flow-rate of the liquid. The width of a zone with Gaussian profile is equal to $4\sigma_t$ at the base.

The overall zone broadening is given by the contributions from all parts of the apparatus and it holds that the variances of systems connected in series are additive

$$\sigma_{t,tot}^2 = \sum_i \sigma_{t,i}^2. \tag{5.52}$$

The main sources of broadening are the solute dispersion in the connecting tubes, the effective working volume of the detector and the dynamics of the sensor, of the electronic circuitry and of the recording device.

For dispersion in a straight, narrow and long tube with radius r and length l,

$$\sigma_t^2 = (t_R r^2)/(24\,D), \tag{5.53}$$

or

$$\sigma_V^2 = (\pi r^4 l w)/(24\,D), \tag{5.54}$$

where t_R is the residence time of the solute in the tube and D is its diffusion coefficient. The dispersion is smaller in coiled tubes, owing to secondary flow of the solution. Equations (5.53) and (5.54) are reasonably accurate only for long residence times and narrower tubes, otherwise the values predicted are too large.

The contribution from the detector effective volume can be approximately expressed as

$$\sigma_V = k V_c, \tag{5.55}$$

where V_c is the cell volume and k is a proportionality constant with a value of 0.3 to 1.0.

The broadening caused by the electronics is constant in time and can often be predicted from the amplifier time constants, i.e. $\sigma_t = C$ or $\sigma_V = wC$.

Therefore, the total volume standard deviation is given by the equation,

$$\sigma_V^2 = \pi r^4 l w/24\,D + k^2 V_c^2 + w^2 C^2. \tag{5.56}$$

To suppress the broadening, the flow-rate should be low, the connecting tubes short and narrow, the detector volume small and the response of the electronic circuitry fast. (For a more detailed discussion and references to the original literature, see [114, 136, 150].)

In addition to these principal requirements, there are further characteristics that affect the performance and applicability of analysers and detectors. The signal should depend as little as possible on the liquid flow-rate and temperature. With continuous monitors, the sensor should be selective for the test component. On the other hand, with chromatographic detectors, the sensor should respond to all eluted components with the same sensitivity.

ISEs are well suited for flow measurements because the instrumentation and signal handling are simple, the measurement is almost independent of the liquid flow-rate, the linear dynamic range is broad, the temperature dependence is not very pronounced and the measurement is selective (the selectivity is, however, a drawback in applications to chromatography). The experimental conditions are readily adjusted and often only consist of ionic strength and pH maintenance. ISEs with solid membranes usually exhibit better performance than liquid membrane electrodes and gas probes, because their response is faster and they are mechanically stronger. The most difficult problem is passivation of the electrodes in some media, for example, biological fluids or surface and waste waters.

With some ISEs, measurement in flowing solutions leads to a substantial improvement in the value of the determination limit (down to 10^{-8} to 10^{-9} mol l^{-1}; see section 5.1.2).

5.3.1 Continuous monitoring

The simplest method of continuous monitoring involves placing an ISE and a reference electrode directly in the test liquid stream. However, this method can rarely be used because some sample pretreatment (filtering off solid particles, adjustment of pH and ionic strength, or masking) is usually necessary. Therefore, a small part of the liquid is drawn from the stream, any solid particles present are filtered off, the solution is mixed with the reagents and then fed to the detector cell.

The use of an ISE with a reference electrode of the second kind has certain drawbacks, such as the existence of a poorly reproducible liquid–junction potential, frequent blocking of the reference electrode liquid bridge and sometimes difficult calibration under the flow conditions. Therefore, differential techniques (see section 5.2.4) are usually preferred. Two identical ISEs can be located in various places in the flow system [102] (see fig. 5.4). It is then possible to measure the potential differences, $E_1 - E_2, E_1 - E_3$, and $E_3 - E_2$. When the reagent contains a known, constant concentration of the test substance, the $E_1 - E_2$ potential difference directly yields the sample concentration because electrode 1 functions as a reference electrode. Difference $E_1 - E_2$ yields the sample concentration by the analyte addition method and difference $E_3 - E_2$ by the known addition method. If the reagent contains a complexing agent, the sample concentration is obtained by the known subtraction method when difference $E_3 - E_2$ is measured. Substances for which an ISE is not available can be determined when they are complexed or precipitated by a reagent sensed by an ISE; the $E_3 - E_1$ difference then yields the sample concentration by the analyte subtraction method. Systems containing an ISE in position 3 have an additional advantage in that calibration with standard solution is not necessary. The test substance is pumped through the branch containing the ISE and flow

Fig. 5.4. Various positions of ISE's in a simple flow-through system [105].

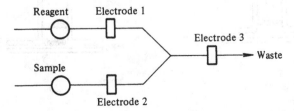

in the branch that does not contain an ISE is stopped; any difference in the ISE potentials is then compensated by a suitable external voltage.

Usually it is impossible to use several different ISEs in a single flow-through cell, because each electrode requires different sample pretreatment. However, it is possible to separate the sample stream into several parts that are fed to flow-through cells containing individual ISEs and to handle the signals by a minicomputer [140, 167]. Up to five ISEs were monitored in this way, using the standard addition method and the rigorous least squares fit for each ISE [167]. The data were sampled with a resolution of ± 0.002 mV by multi-plexing the five ISEs within 1 second and the digitized signals were displayed in real time, with a relative error of 0.4 to 3.3% and confidence intervals from ± 0.4 to $\pm 1.6\%$.

In common industrial and laboratory continuous analyzers, the requirements on the magnitude of the cell are not as rigorous as with chromatographic, CFA and FIA detectors and thus the construction of the detector can be quite simple. The principal requirements are a sufficiently high and reproducible flow-rate in the vicinity of the ISE membrane, good electrical connection between the ISE and the reference electrode by the flowing liquid and, when a differential circuit is not used, placement of the reference electrode after the ISE to avoid possible interference from solution leaking from the reference electrode.

The simplest flow-through cells are actually caps fitted at the end of the ISE and the reference electrode and connected by a tube (see fig. 5.5). Some researchers [80, 92] claim that stirring the solution inside the cap by a rod driven by an external magnet (fig. 5.5b) accelerates the ISE response and improves the potential stability (see also section 5.1.2). The solution can also be led

Fig. 5.5. Thin-layer flow-through cells [9]: (a0 without stirring; (b) with a magnetic stirrer.

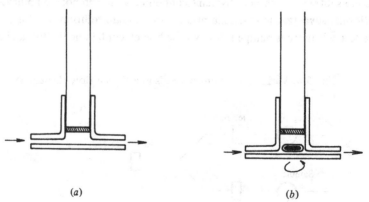

(a)   (b)

through a channel drilled in the ISE membrane; however, bubbles in the test
solution are readily trapped in the channel. A flow-through electrode with
a liquid membrane based on this principle [111] is shown in fig. 5.6. Industrial
analysers, for which a sufficient amount of sample solution is available and the
time constant is not critical, employ simple plastic cells of the type shown in
fig. 5.7.

An interesting cell construction, described in [128], is suitable especially for
clinical analysers, which limits the dead space and thus optimizes solution flow
(fig. 5.8). The ISE is placed in a channel and the elevation in the channel
opposite the electrode ensures that the flow-rate is highest at the ISE membrane.

Fig. 5.6. A flow-through electrode system for liquid-membrane ISEs
[111]: 1 – reference electrode; 2 – hole through which the sample
solution flows; 3 – liquid ion-exchanger reservoir; 4 – triangular piece of
a frit soaked with the liquid ion-exchanger.

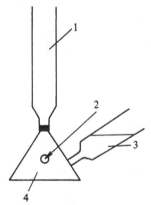

Fig. 5.7. A typical flow-through cell for an industrial analyser [9]:
1 – ISE; 2 – reference electrode; 3 – space for thermometer.

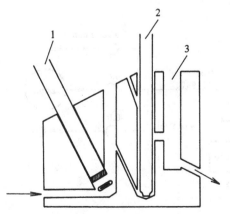

A frequently encountered difficulty when working with peristaltic pumps is the formation of streaming potentials as a result of flow pulsations. Streaming potentials are caused by electrokinetic effects and their value increases with increasing flow-rate and tube length, and with decreasing tube cross-section and solution electrical conductance. They may attain values of several tens of millivolts and thus cause large measuring errors. Therefore, the experimental conditions must be optimized in this respect. The streaming potential value can be substantially suppressed when a grounding electrode is placed in the liquid stream [154, 156].

Fig. 5.8. A flow-through cell with optimized liquid flow [128].

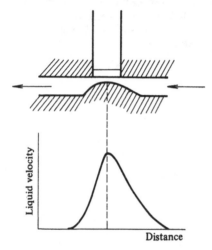

Fig. 5.9. Free cyanide monitor [35].

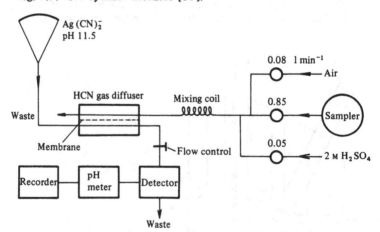

The above principles have been employed in the construction of a monitor of gaseous inorganic fluorine and chlorine compounds after absorption of the gases in a solution [13], a monitor of free cyanide [35] (fig. 5.9), and a monitor of sulphate using the reaction of the sulphate with a Pb^{2+} standard solution and, following the decrease in the Pb^{2+} concentration, with a Pb^{2+} ISE [152] (fig. 5.10). Analysers for clinical purposes have been designed [129–131] and even a bedside analyser for monitoring Na^+, K^+, Ca^{2+} and β-D-glucose in patients blood [127] or a blood potassium analyser for use during open-heart surgery [109]. A computer-controlled interference correction has been proposed [44], in which the standards are mixed to match the electrode potential obtained in the test solution. A simple calibration in flow systems [61] involves dilution of the standard solution and monitoring of the ISE potential as a function of the diluent volume and dilution time.

5.3.2 Titration techniques in flowing solutions

Titration techniques can be applied both to continuous monitoring and to analyses of discrete samples in flow-through systems. Generally it holds that the equivalence point for the reaction

$$aA + bB \rightarrow A_aB_b \tag{5.57}$$

is reached when it holds that

Fig. 5.10. Sulphate Monitor [152].

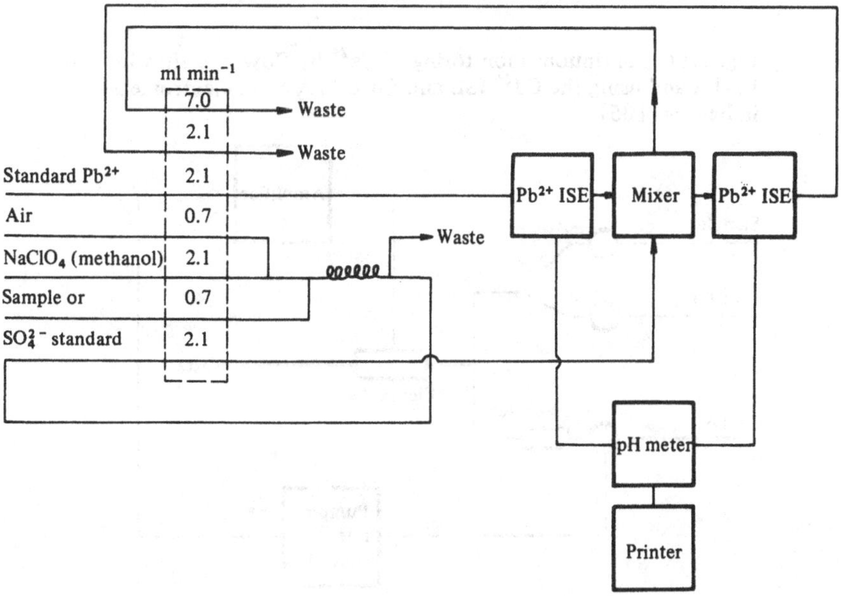

$$ac_A V_A = bc_B V_B \tag{5.58}$$

where c_A and c_B are the concentrations of A and B, respectively, and V_A and V_B are their respective volumes. When the solutions are continuously mixed, then the volumes can be replaced by volume flow-rates, w_A and w_B. Therefore, one of the parameters is varied (a concentration or a flow-rate) and the other three are held constant, until the equivalence point is attained.

In the original work [12], the reagent flow-rate was varied. An example of such a titration is the determination of calcium by titration with EDTA using the Cd^{2+} ISE and the Cd–EDTA complex as an electrometric indicator [105] (fig. 5.11). Another approach is variation of the titrant concentration while the titrant flow-rate is maintained constant [40]. When the titrant is pumped at a constant rate into a vessel with a solvent in which the mixture is stirred and led to the measuring cell, a linear concentration gradient is obtained. If the titrant concentration is $c_{0,T}$ and the pumping rate into the vessel with the solvent is $v_{0,T}$, c_T being the instantaneous concentration of the solution and v_T the rate of the pumping of the solution into the flow-through system, then c_T is given by

$$c_T = c_{0,T} + (c_T^{init} - c_{0,T}) \left[\frac{(v_{0,T} - v_T)t + v_T}{V_T^{init}} \right] \exp \frac{v_{0,T}}{(v_T - v_{0,T})}, \tag{5.59}$$

where c_T^{init} is the concentration and V_T^{init} is the volume of the solution formed at time $t = 0$. For $v_T = 2v_{0,T}$, (5.59) is simplified to

$$c_T = c_{0,T} + (c_T^{init} - c_{0,T})/V_T^{init} [V_T^{init} - v_{0,T}t]e. \tag{5.60}$$

Fig. 5.11. Continuous monitoring of Ca^{2+} by flow-rate titration with EDTA and using the Cd^{2+} ISE and Cd EDTA as an electrometric indicator [105].

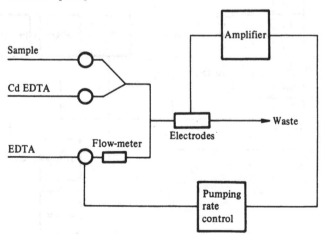

Hence, c_T is a linear function of time. An example of such a titration is the determination of sulphide with an Hg^{2+} solution, using S^{2-} ISE [40] (fig. 5.12).

In an effort to improve the accuracy of the determination of the end-point, the triangle-programmed titration was devised [97]. The sample with concentration c_x flows at a constant rate of w_s. The titrant is introduced at programmed volume flow-rate V_R, the change in V_R following the pattern of an isosceles triangle (fig. 5.13a). The time of the whole program is 2τ, for $t \leqslant \tau$, $V_R = nt$; for $t > \tau$, $V_R = (2\tau - t)n$; n and τ are constants. The maximum titrant mass flow should exceed the sample mass flow ($V_s = c_x w_s$). Two connected titration curves are obtained (fig. 5.13b), the time elapsed between the two equivalence points for the titration reaction

$$aA + bB \rightarrow dP_1 + gP_2 \tag{5.61}$$

Fig. 5.12. Continuous concentration gradient titration of sulphide with mercuric nitrate, using an S^{2-} ISE [40].

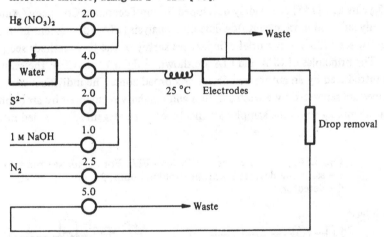

Fig. 5.13. A triangle-programmed potentiometric titration: (a) the titrant addition program; (b) the titration curves. For details see the text.

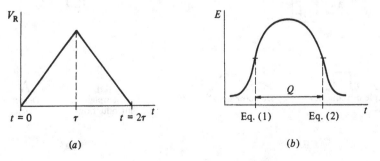

(a)　　　　　　　　　　　　　(b)

being

$$Q = t_{eq(2)} - t_{eq(1)} = 2\tau - 2ac_x w_s/b \tag{5.62}$$

The sample concentration can be calculated from (5.62). The handling of the results is more complicated than in normal titrations because two end-points must be determined and thus a desk-top calculator was recommended for the purpose [100]. From the point of view of the accuracy and precision, it is advantageous to generate the titrants coulometrically [98, 99].

5.3.3 Analyses of discrete samples in flow systems and chromatographic detection

Routine analyses of large numbers of similar samples can readily be automated and the sample throughput considerably increased (sometimes up to about 200 samples per hour) by carrying out the analyses in a continuously flowing medium. At present there are two basic approaches to the problem, the older technique of continuous-flow analysis (CFA) introduced more than 25 years ago [145] and widely developed by the Technicon Company (Auto-Analyzer), and more recent flow-injection analysis (FIA; for a recent literature review see [123]). For a brief comparative survey of the two methods see [148].

The principles of CFA and FIA are shown in fig. 5.14. In CFA, the sample is introduced by an automated device as a broad zone. The individual sample zones are separated by a wash solution and regularly segmented by air bubbles. In the mixing spiral the samples are incubated with reagents and are led into

Fig. 5.14. The principles of CFA and FIA. For details see the text.
1 – sampling device; 2 – proportioning pump; 3 – mixing spiral, 4 – detector.

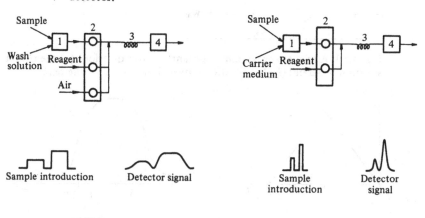

CFA   FIA

a detector on completion of the reaction (i.e. at a steady-state). If required by
the detection technique, the air bubbles are removed before the detector. The
detector signal then has the form of integrated Gaussian curves with flat tops.
In FIA the samples are injected through a septum or by a sampling valve as
narrow zones into a continuously flowing carrier solution, are mixed with
reagents and led in a non-steady-state into a detector. The detector signal contains
relatively narrow Gaussian peaks with sharp maxima. Both in CFA and FIA the
peak height is proportional to the sample concentration and the peak area to
the sample amount. The injection technique developed by Pungor and coworkers
(see [117]) is actually an intermediate between CFA and FIA. The apparatus
and procedure are similar to FIA but a mixing chamber is included in which the
system attains steady-state before entering the detector.

The relative merits and drawbacks of CFA and FIA may be summarized as
follows:

(1) FIA is instrumentally simpler than CFA, mainly because it does not
    require introduction and removal of air bubbles.
(2) The sample throughput in FIA is about twice that in CFA and the
    sample and reagent consumption are about half of those in CFA.
(3) The detection sensitivity is higher in CFA because the samples are not
    diluted by a carrier medium.
(4) The precision and reliability of the results are better in CFA and depend
    less on the fluctuations in the sample size because of a better defined
    concentration profile in the detector (steady-state) and because it is
    possible to average the data over the whole flat top of the curve; this of
    course requires more complex mathematical data treatment.
(5) Because of segmentation with air bubbles, the diffusion broadening of
    the sample zones is much smaller with CFA, leading to a smaller danger
    of carryover between the samples. It is also a great advantage when the
    determination requires a longer incubation time, especially when preli-
    minary separations are involved, as in dialysis, solvent extraction, column
    separations, evaporation to dryness, etc.
(6) An advantage of FIA is the possibility of using carrier liquids that
    contain all necessary reagents or of generating the reagents coulometri-
    cally in the carrier liquid.

Hence FIA is especially suitable for rapid analyses of simple systems whereas
CFA is advantageous in more complicated analyses. The theory of CFA and FIA
is based on the considerations on zone broadening mentioned above (5.51) to
(5.56). It still contains many serious simplifications and thus all conclusions (see
[123, 148]) are semi-empirical. It has been shown [148] that, to attain sufficient

separation of the sample zones, the distance between the zones must equal $8\sigma_t$ for CFA and $4\sigma_t$ for FIA. Therefore, the maximum analysis frequency, f (sample number per hour), is

$$f = 3600/4\sigma_t = 900/\sigma_t \quad \text{(FIA)} \tag{5.63}$$

and

$$f = 3600/8\sigma_t = 450/\sigma_t \quad \text{(CFA)}. \tag{5.64}$$

In view of the zone dispersion, the ISE cells depicted in figs. 5.5 and 5.7 obviously have too large volumes. For CFA and FIA, either small volume thin-layer cells (fig. 5.15) or wall-jet cells (fig. 5.16) are suitable. The effective cell volume should not exceed about 20 microlitres. With a sample size of several tens of microlitres, the analytical frequency then varies between about 40 and 200 samples per hour.

CFA with ISE detection has been applied to determinations with enzyme electrodes [81]. It has been shown [115] that potentiometric detection with ISEs is superior to colorimetric detection in its simpler methodology, high selectivity, a wide linear dynamic range, and insensitivity to solution colouration

Fig. 5.15. A thin-layer cell with an ISE for FIA [125].

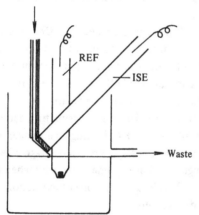

Fig. 5.16. A wall-jet cell with an ISE [9].

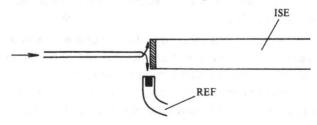

and small amounts of precipitates, but often suffers from the formation of streaming potentials. When using the Technicon SMAC analyser controlled by a dedicated computer [1], sodium and potassium ions can be determined in serum using ISEs with a frequency of up to 200 samples per hour [118] (the segmenting air bubbles are not removed) and with a precision better than that of flame photometry [165].

Examples of the use of FIA with ISE detection involve the determination of nitrate and total nitrogen in environmental samples [48, 49, 125, 166], potassium, sodium [125], calcium [51] and urea [124] in serum or major nutrients in fertilizers [73]. An interesting combination of an ISFET sensor with the FIA principle [52] is shown in fig. 5.17. This is a simultaneous determination of potassium, calcium and pH in serum during dialysis on an artificial kidney. In the bypass position, the carrier solution flows through the bypass loop and across the ISFET. The sample is injected into the sampling valve and is introduced into the carrier solution. The bypass loop has a high hydrodynamic resistance and thus the solution proceeds to the detector. The reference electrode is always immersed only in the carrier solution and is electrically connected with the ISFET through the solution. The apparatus is regularly calibrated by K^+, Ca^{2+} and pH standard solutions.

ISEs have many attractive features for use as detectors in high-performance liquid chromatography (HPLC) (see [149, 150]). It is easy to design a cell with an extremely small internal volume required by HPLC, the measuring technique and signal handling are simple and the detector signal is virtually independent of the effluent flow-rate. However, the ISE response is too selective for most chromatographic applications and usually also too slow, especially at low solute concentrations. Moreover, the necessity of using mobile phases with a relatively high and constant ionic strength limits the use of ISEs almost exclusively to inorganic systems and prevents the technique of gradient elution. These problems can be somewhat alleviated by using auxiliary reactions in a cell preceding the detector, but the system then becomes too complex. For these reasons, detectors with ISEs have had difficulties in competing with other HPLC detectors and have only rarely been used, chiefly in gel and ion-exchange chromatography [31, 32, 34, 95, 135]. A non-selective differential potentiometric detector for ionized substances [30] bears greater promise for HPLC. It consists of two chambers separated by an ion-exchanger membrane, the effluent passing through one chamber and the eluent through the other. The potential difference at the membrane is monitored by two reference electrodes placed in the chambers; the detection limit is about 1 nanomole, the linear dynamic range 2 to 3 concentration decades, the relative standard deviation of the measurement above 3% and the time constant 2 seconds.

ISEs have also been used as detectors in gas chromatography, with post-column derivatization. Organic substances containing fluorine were hydrogenated, the hydrogen fluoride formed was absorbed and detected by a fluoride ISE [69], sulphur-containing substances were either catalytically hydrogenolysed with measurement of the sulphide formed [71] or absorbed in an $AgNO_3$ solution with monitoring of the decrease in the Ag^+ concentration [72] and bromine- and chlorine-containing compounds were hydrogenated and detected, after absorption of HBr and HCl, by the appropriate ISEs [70]. These detectors are sensitive, with low noise and stable baseline, but the system is rather complicated and its response is slow (up to 10 seconds at a concentration of 10^{-4} M), which causes broadening of the elution curves.

5.4 Specific features of ISE application in clinical analysis

It has been correctly pointed out in a recent review [10] that there is very extensive literature on the possible use of ISEs in clinical analysis, but that

Fig. 5.17. (*a*) A combination of FIA with ISFET detection [52]. For a description, see the text. (*b*) A recording of simultaneous determination of potassium, calcium and pH in serum by FIA with ISFET detection [52]. A – potassium standard; B – calcium plus pH standard; 0 – baseline values in the carrier solution (physiological solution).

(*a*)

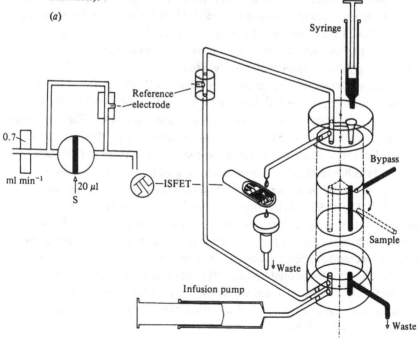

(b)

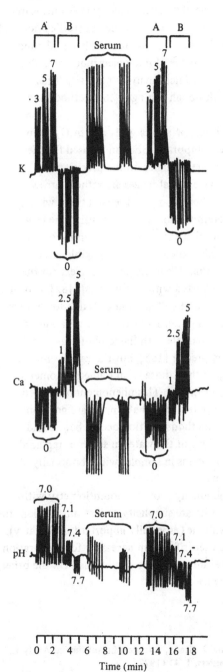

the actual routine use in clinical practice lags behind. This is caused by the fact that biological materials are very complicated matrices and that for measurements *in vivo* it is necessary to fulfill many conditions, such as sterilization and non-toxicity of the device, coping with large and irreproducible liquid–junction potentials, extremely high electrical resistance of microelectrodes, etc. Moreover, the great responsibility involved in clinical analyses naturally leads to conservatism and to requirements of thorough testing of the methods and instruments before their routine use.

Among inorganic components, pH, pK, pNa, pCa, pCl_2 and pCO_2 are of the greatest importance for medicine. Contemporary ISEs can be used for the measurement of these parameters, but the demands on the precision and accuracy are very high because the physiological ranges are rather narrow.

The results obtained with ISEs have been compared several times with those of other methods. When the determination of calcium using the Orion SS-20 analyser was tested, it was found that the results in heparinized whole blood and serum were sufficiently precise and subject to negligible interference from K^+ and Mg^{2+} ([82]), but that it is necessary to correct for the sodium error, as the ionic strength is adjusted with a sodium salt [82], and that a systematic error appears in the presence of colloids and cells due to complexation and variations in the liquid–junction potential [76]. Determination of sodium and potassium with ISEs is comparable with flame photometric estimation [39, 113, 116] or is even more precise [165], but the values obtained with ISEs in serum are somewhat higher than those from flame photometry and most others methods [3, 25, 27, 113, 116]. This phenomenon is called pseudohyponatremia. It is caused by the fact that the samples are not diluted in ISE measurement, whereas in other methods dilution occurs before and during the measurement. On dilution, part of the water in serum is replaced by lipids and partially soluble serum proteins in samples with abnormally increased level of lipids and/or proteins.

The main advantages that should gradually promote potentiometry with ISEs to routine use in clinical laboratories are simplicity of instrumentation, the possibility of decreasing the sample volume (especially important in pediatry), and the possibility of avoiding tedious centrifugation necessary for preparation of plasma and serum (significant under intensive care conditions). On the other hand, work with ISEs requires experience and skill.

References for chapter 5

1 H. Amar, S. Barton, D. Dubac and J. Grady, Advances in Automated Analysis, *Technicon International Congress*, 1, 41 (1972).
2 T. Anfält and D. Jagner, *Anal. Chim. Acta* 57, 165 (1971).

3 W. Annan, N. A. Kirwan and W. S. Robertson, *Clin. Chem.* **25**, 1865 (1979).
4 J. M. Ariano and W. F. Gutknecht, *Anal. Chem.* **48**, 281 (1976).
5 Y. Asano and S. Ito, *Nippon Kagaku Kaishi* 1498 (1980).
6 O. Åström, *Anal. Chim. Acta* **88**, 17 (1977).
7 O. Åström, *Anal. Chim. Acta* **97**, 259 (1978).
8 H. D. Axelrod and S. G. Hansen, *Anal. Chem.* **47**, 2460 (1975).
9 P. L. Bailey, *Ion-Sel. El. Rev.* **1**, 81 (1979).
10 D. M. Band and T. Treasure, Ion-Selective Electrodes in Medicine and Medical Research, in *Ion-Selective Electrode Methodology* (ed. A. K. Covington), CRC Press, Boca Raton 1979.
11 E. W. Baumann and R. M. Wallace, *Anal. Chem.* **41**, 2072 (1969).
12 W. J. Blaedel and R. H. Lessing, *Anal. Chem.* **36**, 1617 (1964).
13 G. Blažević, M. Boehner, E. Schenbeck, *Fresenius Z. Anal. Chem.* **298**, 12 (1979).
14 R. Blum and H. M. Fog, *J. Electroanal. Chem.* **34**, 485 (1972).
15 M. J. D. Brand and G. A. Rechnitz, *Anal. Chem.* **42**, 616 (1970).
16 M. J. D. Brand and G. A. Rechnitz, *Anal. Chem.* **42**, 1172 (1970).
17 L. G. Bruton, *Anal. Chem.* **43**, 579 (1971).
18 J. Buffle, *Anal. Chim. Acta* **59**, 439 (1972).
19 J. Buffle, N. Parthasarathy and D. Monnier, *Chimia* **25**, 224 (1971).
20 J. Buffle, N. Parthasarathy and D. Monnier, *Anal. Chim. Acta* **59**, 427 (1972).
21 P. R. Burton, Instrumentation for Ion-Selective Electrodes, in *Ion-Selective Electrode Methodology* (ed. A. K. Covington), CRC Press, Boca Raton 1979.
22 K. Cammann, *Das Arbeiten mit Ionselektiven Elektroden*, Springer-Verlag, Berlin (1973).
23 P. W. Carr, *Anal. Chem.* **43**, 425 (1971).
24 P. W. Carr, *Anal. Chem.* **44**, 452 (1972).
25 R. L. Coleman, *Clin. Chem.* **25**, 1865 (1979).
26 A. K. Covington, Reference Electrodes, in *Ion-Selective Electrodes* (ed. R. A. Durst) NBS No. 314 Special Publ., Washington (1969).
27 J. D. Czaban and A. D. Cormier, *Clin. Chem.* **26**, 1921 (1980).
28 T. Damokos and J. Havas, *Hung. Sci. Instr.* **36**, 7 (1976).
29 L. A. Dee, H. H. Martens, C. I. Merrill, J. T. Nakamura and F. C. Jaye, *Anal. Chem.* **45**, 1477 (1973).
30 R. S. Deelder, H. A. J. Linssen and J. G. Koen, *J. Chromatogr.* **203**, 153 (1981).
31 T. Deguchi, A. Hisanaga and H. Nagai, *J. Chromatogr.* **133**, 173 (1977).
32 T. Deguchi, T. Kuma and H. Nagai, *J. Chromatogr.* **152**, 349 (1978).
33 J. Doležal, P. Povondra and Z. Šulcek, *Decomposition Techniques in Inorganic Analysis*, Iliffe, London (1969).
34 R. C. Dorey, III, Univ. Georgia Microfilms, No. 8017162, Athens, USA 1980.
35 R. A. Durst, *Anal. Lett.* **10**, 961 (1977).
36 R. A. Durst, *Clin. Chim. Acta* **80**, 225 (1977).
37 R. A. Durst, E. L. May and J. K. Taylor, *Anal. Chem.* **40**, 977 (1968).
38 R. A. Durst and J. K. Taylor, *Anal. Chem.* **39**, 1374 (1967).
39 P. Fievet, A. Truchaud, J. Hersant and G. Glikmanas, *Clin. Chem.* **26**, 138 (1980).
40 B. Fleet and A. Y. W. Ho, *Anal. Chem.* **46**, 9 (1974).
41 M. S. Frant and J. W. Ross, Jr., *Anal. Chem.* **40**, 1169 (1968).
42 J. W. Frazer, A. M. Kray, W. Selig and R. Lim, *Anal. Chem.* **47**, 869 (1975).
43 J. W. Frazer, W. Selig and L. P. Rigdon, *Anal. Chem.* **49**, 1250 (1977).
44 P. D. Gaarenstroom, J. C. English, S. P. Perone and J. W. Bixler, *Anal. Chem.* **50**, 811 (1978).
45 G. Gran, *Acta Chem. Scand.* **4**, 559 (1950).
46 G. Gran, *Analyst* **77**, 661 (1952).

47 J. L. Guth and R. Wey, *Bull. Soc. Fr. Miner. Cristall.* **92**, 105 (1969).
48 E. H. Hansen, A. K. Ghose and J. Ruzicka, *Analyst* **102**, 705 (1977).
49 E. H. Hansen, F. J. Krug, A. K. Ghose and J. Ruzicka, *Analyst* **102**, 714 (1977).
50 E. H. Hansen, C. G. Lamm and J. Růžicka, *Anal. Chim. Acta* **59**, 403 (1972).
51 E. H. Hansen, J. Ruzicka and A. K. Ghose, *Anal. Chim. Acta* **100**, 151 (1978).
52 J. Harrow, J. Janata, R. L. Stephenson and W. J. Kolff, private communication.
53 J. E. Harwood, *Water Res.* **3**, 273 (1969).
54 S. J. Haynes and A. H. Clark, *Econom. Geol.* **67**, 378 (1972).
55 E. Heckel and P. F. Marsh, *Anal. Chem.* **44**, 2347 (1972).
56 G. Horvai, L. Domokos and E. Pungor, *Z. Anal. Chem.* **292**, 132 (1978).
57 G. Horvai, L. Domokos and E. Pungor, *Magy. Kém. Foly.* **84**, 481 (1978).
58 G. Horvai and E. Pungor, *Anal. Chim. Acta* **113**, 287 (1980).
59 G. Horvai and E. Pungor, *Anal. Chim. Acta* **113**, 295 (1980).
60 G. Korvai, K. Tóth and E. Pungor, *Magy. Kém. Foly.* **84**, 483 (1978).
61 G. Horvai, K. Tóth and E. Pungor, *Anal. Chim. Acta* **82**, 45 (1976).
62 G. Horvai, K. Tóth and E. Pungor, *Anal. Chim. Acta* **107**, 101 (1979).
63 F. Ingmann and E. Still, *Talanta* **13**, 1431 (1966).
64 A. Ivaska, *Talanta* **21**, 377, 387 (1974).
65 H. Jenny, T. R. Nielsen, N. T. Coleman and D. E. Williams, *Science* **112**, 164 (1950).
66 G. Johansson and W. Backén, *Anal. Chim. Acta* **69**, 415 (1974).
67 B. Karlberg, *Anal. Chem.* **43**, 1911 (1971).
68 R. N. Khuri, Glass Microelectrodes and their Uses in Biological Systems, in *Glass Electrodes for Hydrogen and other Cations* (ed. G. Eisenman), M. Dekker, New York 1967.
69 T. Kohima, M. Ichise and Y. Seo, *Talanta* **19**, 539 (1972).
70 T. Kojima, M. Ichise and Y. Seo, *Anal. Chim. Acta* **101**, 273 (1978).
71 T. Kojima, Y. Seo and J. Sato, *Bunseki Kagaku* **23**, 1389 (1974).
72 T. Kojima, Y. Seo and J. Sato, *Bunseki Kagaku* **24**, 772 (1975).
73 M. Koshino, *Bunseki Kagaku* **11**, 803 (1978).
74 J. A. Krueger, *Anal. Chem.* **46**, 1338 (1974).
75 J. Langmaier, R. Kalvoda and K. Štulík, *Anal. Chim. Acta*, in press.
76 L. Larsson and S. Öhman, *Clin. Chem.* **24**, 731 (1978).
77 W. Leithe, *Chem.-Ing. – Tech. Z.* **36**, 112 (1964).
78 T. K. Li and J. T. Piechocki, *Clin. Chem.* **17**, 411 (1971).
79 A. Liberti and M. Mascini, *Anal. Chem.* **41**, 676 (1969).
80 W. A. Lingerak, F. Bakker and J. Slanina, in *Proceedings of the Conference on Ion-Selective Electrodes, Budapest 1977* (ed. E. Pungor), Akadémiai Kiadó, Budapest 1978, p. 453.
81 R. A. Llenado and G. A. Rechnitz, *Anal. Chem.* **45**, 826 (1973).
82 S. Madsen and K. Ølgaard, *Clin. Chem.* **23**, 690 (1977).
83 H. V. Malmstadt and J. D. Winefordner, *Anal. Chim. Acta* **20**, 283 (1959).
84 S. E. Manahan, *Anal. Chem.* **42**, 128 (1970).
85 E. A. Mangubat, T. R. Hinds and F. F. Vincenzi, *Clin. Chem.* **24**, 635 (1978).
86 C. R. Martin and H. Freiser, *Anal. Chem.* **51**, 803 (1979).
87 M. Mascini, *Anal. Chim. Acta* **85**, 287 (1976).
88 M. Mascini, *Ion-Sel. El. Rev.* **2**, 17 (1980).
89 L. Meites and J. A. Goldman. *Anal. Chim. Acta* **29**, 472 (1963).
90 L. Meites and J. A. Goldman, *Anal. Chim. Acta* **30**, 18 (1964).
91 L. Meites and T. Meites, *Anal. Chim. Acta* **37**, 1 (1967).
92 J. Mertens, P. Van den Winkel and D. L. Massart, *Anal. Chem.* **48**, 272 (1976).
93 D. Midgley, *Anal. Chem.* **49**, 1211 (1977).

94 D. Midgley, *Analyst* **104**, 248 (1979).
95 M. Midorikawa, Japan Kokai Tokkyo Koho, 8031931; *Chem. Abstr.* **93**, 88089q (1980).
96 E. W. Moore, *J. Clin. Invest.* **49**, 318 (1970).
97 G. Nagy, Z. Fehér, K. Tóth and E. Pungor, *Anal. Chim. Acta* **91**, 87 (1977).
98 G. Nagy, Z. Fehér, K. Tóth and E. Pungor, *Anal. Chim. Acta* **91**, 97 (1977).
99 G. Nagy, Z. Fehér, K. Tóth and E. Pungor, *Magy. Kém. Foly.* **85**, 321 (1979).
100 G. Nagy, Z. Lengyel, Z. Fehér, K. Tóth and E. Pungor, *Anal. Chim. Acta* **101**, 261 (1978).
101 *Orion Newsletter* **I**, 27 (1969).
102 *Orion Newsletter* **II**, 1–34 (1970).
103 *Orion Newsletter* **II**, 1 (1970).
104 *Orion Newsletter* **II**, 5 (1970).
105 *Orion Newsletter* **II**, 7, 8, (1970).
106 *Orion Newsletter* **II**, 26 (1970).
107 *Orion Newsletter* **II**, 49 (1970).
108 *Orion Newsletter* **III**, 35 (1971).
109 H. F. Osswald, R. Asper, W. Dimai and W. Simon, *Clin. Chem.* **25**, 39 (1979).
110 H. Pallman, *Koll. Chem. Beih.* **30**, 334 (1930).
111 D. S. Papastathopoulos, E. P. Diamandis and T. P. Hadjiioannou, *Anal. Chem.* **52**, 2100 (1980).
112 N. Parthasarathy, J. Buffle and D. Monnier, *Anal. Chim. Acta* **59**, 447 (1972).
113 S. Patal and P. O'Gorman, *Clin. Chem.* **24**, 1856 (1978).
114 H. Poppe, *Anal. Chim. Acta* **114**, 59 (1980).
115 J. R. Potts, *7th Technicon International Congress*, (1977), p. 38.
116 C. J. Preusse and C. Fuchs, *J. Clin. Chem. Clin. Biochem.* **17**, 639 (1979).
117 E. Pungor, Z. Fehér, G. Nagy, K. Tóth, G. Horvai and M. Gratzl, *Anal. Chim. Acta* **109**, 1 (1979).
118 K. J. M. Rao, M. H. Pelavin and S. Morgenstern, Advances in Automated Analysis, *Technicon International Congress* **1**, 33 (1972).
119 C. N. Reilley and R. W. Schmid, *Anal. Chem.* **30**, 947 (1958).
120 L. P. Rigdon, G. J. Moody and J. W. Frazer, *Anal. Chem.* **50**, 465 (1978).
121 L. P. Rigdon, C. L. Pomernacki, D. J. Balaban and J. W. Frazer, *Anal. Chim. Acta* **112**, 397 (1979).
122 J. W. Ross, Jr. and M. S. Frant, *Anal. Chem.* **41**, 1900 (1969).
123 J. Růžička and E. H. Hansen, *Flow-Injection Analysis*, J. Wiley, New York, Chichester, Brisbane, Toronto (1981).
124 J. Růžička, E. H. Hansen, A. K. Ghose and H. A. Mottola, *Anal. Chem.* **51**, 199 (1979).
125 J. Růžička, E. H. Hansen and E. A. Zagatto, *Anal. Chim. Acta* **88**, 1 (1977).
126 H. Schäfer, *Z. Anal. Chem.* **268**, 349 (1974).
127 J. G. Schindler, R. Dennhardt and W. Simon, *Chimia* **31**, 404 (1977).
128 J. G. Schindler, M. v. Gülich, H. Maier, G. Stork, W. Schäl, H.-E. Braun, W. Schmid and K.-D-Karaschinski, *Fresenius Z. Anal. Chem.* **301**, 410 (1980).
129 J. G. Schindler, H. Maier, A. Heindland, G. G. Jaros, O. Aziz and C. Steffen, *Med. Progr. Technol.* **7**, 29 (1980).
130 J. G. Schindler, R. G. Schindler and O. Aziz, *J. Clin. Chem. Clin. Biochem.* **16**, 447 (1978).
131 J. G. Schindler, G. Stork, R. Dennhardt, W. Schäl, H.-E. Braun, K.-D-Karaschinski and W. Schmid, *J. Clin. Chem. Clin. Biochem.* **17**, 573 (1979).
132 R. W. Schmid and C. N. Reilley, *J. Am. Chem. Soc.* **78**, 5513 (1956).
133 F. A. Schultz, *Anal. Chem.* **43**, 502 (1971).

134 F. A. Schultz, *Anal. Chem.* **43**, 1523 (1971).
135 F. A. Schultz and D. E. Mathis, *Anal. Chem.* **46**, 2253 (1974).
136 R. P. W. Scott, *Liquid Chromatography Detectors*, Elsevier, Amsterdam (1977).
137 I. Sekerka and J. F. Lechner, *Intern. J. Environ. Anal. Chem.* **2**, 313 (1973).
138 I. Sekerka and J. F. Lechner, *Talanta* **20**, 1167 (1973).
139 I. Sekerka and J. F. Lechner, *Anal. Lett.* **7**, 399 (1974).
140 I. Sekerka and J. F. Lechner, *Anal. Lett.* **7**, 463 (1974).
141 I. Sekerka and J. F. Lechner, *Amer. Lab.* February 1976.
142 W. Selig, *US At. Energy Comm.* UCRL-51167, January 5 (1972).
143 S. Siggia, D. W. Eichlin and R. C. Rheinhart, *Anal. Chem.* **27**, 1745 (1955).
144 R. J. Simpson, Practical Techniques for Ion-Selective Electrodes, in *Ion-Selective Electrode Methodology* (ed. A. K. Covington), CRC Press, Boca Raton 1979.
145 L. T. Skeggs, *Am. J. Clin. Pathol.* **13**, 451 (1957).
146 J. Slanina, F. Bakker, J. J. Möls, J. E. Ordelman and A. G. M. Bruyn-Hes, *Anal. Chim. Acta* **112**, 45 (1979).
147 M. J. Smith and S. E. Manahan, *Anal. Chem.* **45**, 836 (1973).
148 L. R. Snyder, *Anal. Chim. Acta* **114**, 3 (1980).
149 K. Štulík and V. Pacáková, *J. Electroanal. Chem. Interfacial Electrochem.* **129**, 1 (1981).
150 K. Štulík and V. Pacáková, Electrochemical Detection in High-Performance Liquid Chromatography, CRC Crit. Rev. Anal. Chem., CRC Press, Boca Raton, 1983.
151 K. Torrance, *Analyst* **99**, 203 (1974).
152 M. Trojanowicz, *Anal. Chim. Acta* **114**, 293 (1980).
153 Y. Umezawa and S. Fujiwara, *Nippon Kagaku Kaishi*, 1437 (1980).
154 M. Vandeputte, L. Dryon and D. L. Massart, *Anal. Chim. Acta* **91**, 113 (1977).
155 J. M. van der Meer and J. C. Smit, *Anal. Chim. Acta* **83**, 367 (1976).
156 P. van der Winkel, J. Mertens and D. L. Massart, *Anal. Chem.* **46**, 1765 (1974).
157 J. C. van Loon, *Anal. Lett.* **1**, 6 (1968).
158 J. Veselý, D. Weiss and K. Štulík, *Analysis with Ion-Selective Electrodes*, E. Horwood, Chichester (1978).
159 T. R. Virtanen, *Kemia-Kemi* **5**, 460, 614 (1978).
160 T. R. Virtanen, *Kemia-Kemi* **6**, 89 (1979).
161 J. C. Warf, W. D. Cline and R. P. Tavebaugh, *Anal. Chem.* **26**, 342 (1954).
162 T. B. Warner and D. J. Bresan, *Anal. Chim. Acta* **63**, 165 (1973).
163 R. Wawro and G. A. Rechnitz, *Anal. Chem.* **46**, 806 (1974).
164 C. C. Westcott, *Anal. Chim. Acta* **86**, 269 (1976).
165 J. O. Westgard, R. N. Carey, D. H. Feldbruegge and L. M. Jenkins, *Clin. Chem.* **22**, 489 (1976).
166 E. A. Zagatto, B. F. Reis, H. Bergamin and F. J. Krug, *Anal. Chim. Acta* **109**, 45 (1979).
167 J. J. Zipper, B. Fleet and S. P. Perone, *Anal. Chem.* **46**, 2111 (1974).

6

ION-SELECTIVE ELECTRODES WITH SOLID OR GLASSY MEMBRANES

The group of ion-selective electrodes with fixed ion-exchange sites includes systems with various membrane structures. The membranes are either homogeneous (single crystals, pressed pellets, sintered materials) or heterogeneous, set in an inactive skeleton of various polymeric materials. Important electrode materials include silver halides, silver and divalent metal chalcogenides, lanthanum trifluoride and various glassy materials. Here, the latter will be surveyed only briefly, for the sake of completeness.

6.1 Ion-selective electrodes for halide ions

Silver halides have the character of solid electrolytes, where the silver ion acts as the charge carrier (see [125, 204, 266] for AgCl) which moves according to the Frenkel mechanism in the crystal. This type of transport is depicted schematically in fig. 6.1. As the halide ions are located in fixed sites, no diffusion potential is formed within the membrane and (3.4.9) to (3.4.13) are valid for the membrane potential. As mentioned in chapter 3, they can be used for determining either halide ions or silver.

Work with silver halide electrodes is complicated by two specific problems; (a) the effect of nonstoichiometric defects on the membrane behaviour and (b) the effect of the character of the internal contact in the all solid-state electrodes. The effect of excess silver [264] appears both in the dependence of the ISE potential on the determinand activity and in the detection limit. The basic concept in the simplified procedure of Morf et al. [264] depends on the fact that the concentration of silver ions in the vicinity of the membrane is given (for selectivity for silver ions) by the sum of the concentrations of silver ions added to the solution, $[Ag^+]_{added}$, the concentration produced by dissolution of the membrane, $[Ag^+]_{sol}$, and the concentration transferred to the solution from defect silver atoms, $[Ag^+]_{def}$:

$$[Ag^+]_s = [Ag^+]_{added} + [Ag^+]_{sol} + [Ag^+]_{def}. \tag{6.1.1}$$

The solubility product P_{AgX} depends on the overall silver ion concentration at the membrane surface and on the concentration of halide ions resulting from dissolution of the membrane, $[X^-]_{sol}$,

$$P_{AgX} = [Ag^+]_s[X^-]_{sol} = [Ag^+]_s\{[Ag^+]_s - [Ag^+]_{added} - [Ag^+]_{def}\}$$

(6.1.2)

from which it follows that

$$[Ag^+]_s = \tfrac{1}{2}\{[Ag^+]_{added} + [Ag^+]_{def}$$
$$+ [([Ag^+]_{added} + [Ag^+]_{def})^2 + 4P_{AgX}]^{1/2}\}.$$

(6.1.3)

For $[Ag^+]_{def}^2 \gg 4P_{AgX}$, the ISE potential (for a dilute solution) is given by the relationship

$$E_{ISE} = \text{const} + (RT/F)\ln[Ag^+]_s$$
$$= \text{const} + (RT/F)\ln([Ag^+]_{added} + [Ag^+]_{def}).$$

(6.1.4)

The detection limit is given by quantity $[Ag^+]_{def}$. In the opposite limiting case, $[Ag^+]_{def}^2 \ll 4P_{AgX}$, the ISE potential depends on $[Ag^+]_{added}$ in the usual manner;

$$E_{ISE} = \text{const} + (RT/F)\ln\{\tfrac{1}{2}[Ag^+]_{added}$$
$$+ ([Ag^+]_{added}^2 + 4P_{AgX})^{1/2}\},$$

(6.1.5)

with the detection limit $P_{AgX}^{1/2}$.

For anionic response, (6.1.2) must be rearranged to the form

$$P_{AgX} = ([Ag^+]_{def} + [Ag^+]_{sol})[X^-]_s$$
$$= ([Ag^+]_{def} + [X]_s - [X^-]_{added})[X^-]_s.$$

(6.1.6)

Fig. 6.1. Frenkel's scheme for the transport of silver ions in the AgBr crystal lattice. The silver cations are in interstitial positions in the crystal lattice and the same number of unoccupied lattice positions are in the cation part of the lattice.

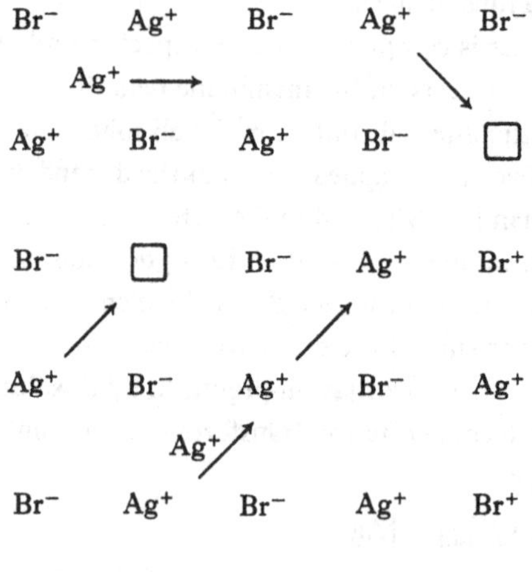

It should be noted that $[Ag^+]_{sol}$ may then be negative. It follows from (6.1.6) for $[X^-]_{added} - [Ag^+]_{def} \gg 2P_{AgX}^{1/2}$ that

$$E_{ISE} = \text{const} + (RT/F) \ln P_{AgX}$$
$$-(RT/F) \ln ([X^-]_{added} - [Ag^+]_{def}) \qquad (6.1.7)$$

while, for $[Ag^+]_{def} - [X^-]_{added} \gg 2P_{AgX}^{1/2}$, it follows that

$$E_{ISE} = \text{const} + (RT/F) \ln ([Ag^+]_{def} - [X^-]_{added}). \qquad (6.1.8)$$

This theory has been successfully verified experimentally. Buck and Shepard [51] demonstrated that electrodes of the all-solid-state type have a response that is identical to that of similar electrodes of the second kind for response to halide ions and to a silver electrode for response to silver ions, depending on the degree of saturation with silver. This is achieved by soldering a silver contact to the membrane. If however the internal contact material is more noble than silver (platinum, graphite, mercury), the electrode with response to silver ions may attain a potential between the standard potential of a silver electrode $E_{Ag^+/Ag}^0$ and the value

$$E_{Ag^+/Ag}^0 + (RT/F) \ln K_0(AgX),$$

where $K_0(AgX)$ is the formation constant of AgX at activity $\bar{a}(AgX)$ in the membrane material (related to pure AgX) from silver with activity $a(AgX)$ and halide at activity $a(X_2)$,

$$K_0(AgX) = \bar{a}(AgX)/[a(Ag)a(X_2)^{1/2}]. \qquad (6.1.9)$$

In practice, three types of membrane based on silver halides are used. The oldest type is based on silver halide precipitate in a matrix of silicone rubber. The construction and preparation of these electrodes were described in chapter 3.1. The modern version of this electrode, produced by Radelkis, Budapest, is a compromise between the original construction described by Pungor *et al.* [310, 311, 313] and a system with a compact membrane. Electrodes with silver chloride, bromide and iodide are manufactured. According to the manufacturer these electrodes should be soaked before use for 1-2 hours in a dilute solution of the corresponding silver halide. They can be used in a pH region from 2 to 12 and the $dE_{ISE}/d \log [X^-]$ value is approximately 56 mV. These electrodes can be employed for various automatic analytical methods (see chapter 5). They can readily be used in mixtures of alcohol with water, for example up to 90% ethanol and methanol and up to 4% *n*-propanol and isopropanol [196]. In mixtures of acetone-water and dimethylformamide-water, they work reliably only in the presence of a large excess of water [197].

Other types of ISE with silver halides are based on homogeneous membranes [6, 383]. With silver chloride or bromide, a single crystal or membrane from a salt melt can be prepared, while silver iodide membranes are prepared from pressed pellets of polycrystalline material. The properties of these electrodes depend on the method used to prepare the pellets and the original precipitate [413]. For example, if AgI is precipitated from a KI solution by a small excess of $AgNO_3$, the γ-cubic modification of the sphalerite type is formed; by contrast, if a small excess of KI is added to the $AgNO_3$ solution, the β-hexagonal modification with wurtzite structure is formed. As could be expected, the ISE behaviour also differs for these materials. At pressures greater than 2.5×10^9 Pa, the β-modification is converted to the γ-modification, indicating the importance of pressure in pellet formation. Differences in behaviour are especially marked in the effects of stirring (i.e. differences in the ISE potential in stirred and quiescent solutions) and in the electrode sensitivity. The γ-modification is probably preferable, but it slowly changes into the β-modification. Fine pores are formed during this change and the surface conductance of the membrane decreases [134, 413], leading to a change in the overall conductance. The pores in the membrane can then lead to membrane permeability for electrolytes [413, 438]. An unwelcome property of these electrodes is their sensitivity to light, appearing as changes in the ISE potential [77, 413].

An improvement [325] in the properties of the halide ISE, both in increased sensitivity and decreased light sensitivity can be attained by using a mixture of silver halide and Ag_2S, which is far less soluble than any of the silver halides. The other ISE properties do not change. The E_{ISE} versus $\log a_X$ dependence is thus identical with the dependence for a pure silver halide membrane [288]. This system is used in most commercial ISEs.

It is interesting that the conductance of a membrane consisting of a mixture of AgI and Ag_2S is even larger than that of a membrane of Ag_2S alone [413]. The ternary compound Ag_3SI is probably formed [325], and the amount increases with increasing pressure [413].

These types of electrodes for chloride and bromide ions have resistances of less than 30 MΩ, while that of the iodide ISE lies between 1 and 5 MΩ [263]. For a detailed study, see [421–424]. In general, all-solid-state ISEs are used with an internal silver contact. Selectrodes (see p. 60) can also be prepared from a mixture of halide and silver sulphide [328, 329].

Other versions of halide electrodes are the systems Hg_2Cl_2 – HgS [22, 397, 398, 447], Hg_2Br_2 – HgS [165, 398], AgCl – Hg_2Cl_2 [421, 423] and Hg_2Cl_2 – Ag_2S [201]; (see also [245]).

The ISEs described in this section are useful primarily for determination of halide ions by direct potentiometry, where the silver halide in the membrane is identical with the determinand. As follows from the discussion on p. 48 an electrode made of a less soluble silver halide X^- can be used to determine other halide Y^- if the condition

$$a_{X^-} < k^{pot}_{X^-, Y^-} a_{Y^-} \qquad (6.1.10)$$

is fulfilled. Thus the iodide electrode can be used for precise potentiometric titrations of Cl^-, Br^- and I^- [311]. It is necessary to add $Ba(NO_3)_2$ to prevent adsorption of the titrant on the precipitate [68]. The individual halides and rhodanide can be determined using the corresponding ISEs after sample combustion and chromatographic separation [205].

On the other hand, bromide ions interfere in the functioning of the chloride electrode by entering the membrane with formation of $AgCl_x Br_y$, as demonstrated by Sandifer [333] using neutron activation analysis, X-ray diffraction and electron scanning microscopy. A similar unwelcome effect is caused by precipitation of AgI on AgCl [124]. Sulphides, thiosulphates and cyanides, as well as reducing agents, interfere in potentiometric determinations using silver halide ISEs. Thus an ISE is not suitable for the determination of halides in photographic developers. Although all halide ISEs can be used to determine silver ions, the ISE containing $Ag_2 S$ is most suitable [313]. The selectivity coefficients of halide ISEs are listed in the Appendix.

A practically complete survey of the applications of halide ISEs is given in tables in [210, 211, 212, 213]. Consequently, the following discussion will consider only the special properties of the individual halide ISEs and their more important applications.

Chloride ion-selective electrodes The most important region of application is the determination of chlorides in waters, including sea water (for a review, see [167]), in serum [110, 112, 371] (review in [167]) and in soil [151, 219, 341]. The determination of chloride ions in sweat made screening for cystic fibrosis possible in new-born babies (review, [45, 55a, 262]). Br^-, I^- and S^{2-} interfere in the determination of chlorides in phosphate rocks [81]. Sulphite can be determined directly using an electrode with an $Hg_2 Cl_2$ - HgS membrane [398] on the basis of the reaction

$$Hg_2 Cl_2 + 2SO_3^{2-} \rightarrow Hg + Hg(SO_3)_2^{2-} + 2Cl^-. \qquad (6.1.11)$$

The chloride concentration thus corresponds to the sulphite concentration in the test solution. The chloride ISE can be used for various thermodynamic measurements, such as determination of the activity of NaCl [361], also in the pressure range 0.1–60 MPa and temperature of 0–25 °C [217]. This ISE is also suitable for kinetic measurements [32, 202, 215, 368]).

Bromide ion-selective electrodes These electrodes have important applications in the determination of Br^- in waters [437], in blood plasma [70, 308] and in rocks [7, 81]. Br^- ISEs are suitable for studying the oscillating Zhabotinsky reaction [132, 209, 216] and silver-bromide complexes [69], and can also be used in the direct potentiometric determination of SCN⁻ [208].

Iodide ion-selective electrode The iodide electrode has broad application both in the direct determination of iodide ions present in various media as well as for the determination of iodide in various compounds. It is, for example, important in the determination of iodide in milk [44, 64, 218, 382, 442]. This electrode responds to Hg^{2+} ions [150, 306, 439] and can be used for the indirect determination of oxidizing agents that react with iodide, such as IO_3^- [305], IO_4^- [158], Pd(II) [117, 347, 405] and for the determination of the overall oxidant content, for example in the atmosphere [393]. It can also be used to monitor the iodide concentration formed during the reactions of iodide with hydrogen peroxide or perborate, catalyzed by molybdenum, tungsten or vanadium ions, permitting determination of traces of these metals [12, 192, 193, 194, 195]. The permeability of bilayer lipid membranes for iodide can be measured using an I⁻ ISE [348]. For determination of low iodide concentrations, see [171].

Model 97-70 electrode from Orion Research, an electrode for residual chlorine [321], is based on a cell consisting of a platinum electrode which reacts to the I_2/I^- system, and an iodide ISE. The EMF of this cell is given by the relationship

$$E = E_{I_2/I^-} - E_{ISE}$$
$$= E^0_{I/I^-} + (RT/F) \ln([I_2]^{1/2}/[I^-]) - E_{0,ISE} + (RT/F) \ln[I^-]$$
$$= const + (RT/2F) \ln[I_2]. \qquad (6.1.12)$$

In acetate buffer (pH 4), active chloride stoichiometrically oxidizes iodide to iodine, so that this sensor can be used to determine the concentration of active chlorine. This method is useful for the determination of active chlorine in the range, 3–100 ppb.

Scanning electron microscopy has indicated that the deterioration in the electrode behaviour after prolonged contact with a 1 M iodide solution is a result of pore formation [375].

Cyanide ion-selective electrode As demonstrated in chapter 3.4, the dissolution of a halide ISE in the presence of some complexing agents, especially cyanide, can be used for determination of these agents [312, 392, 434]. An iodide ISE can be used as a cyanide ISE. The principal application of this electrode is in the determination of cyanides in waters [60, 126, 281, 336] and in galvanic baths [222, 225].

A flow-through electrode has also been constructed for the determination of cyanide [282]. In some media it is preferable to first separate cyanide by distillation [296]. Formaldehyde, thioglycolic acid, hydroxylamine, pyridine and pyrazolone interfere in the determination of cyanide using an ISE [280]. The effect of dissolution of the surface of an AgI membrane used as a cyanide ISE has been studied using energy dispersion x-ray fluorescence spectroscopy [14a].

The determination of β-glucosidase can be based on the decomposition of amygdaline by this enzyme and measuring the cyanide formed with a CN^- ISE [237]. The enzyme rhodanase catalyzes the reaction $S_2O_3^{2-} + CN^- \rightarrow SCN^- + SO_3^{2-}$. The decrease in the cyanide concentration is then monitored using a cyanide ISE and the rate constant of the reaction is proportional to the rhodanase activity [174, 238].

The thiocyanate ion-selective electrode [21, 247] This has a heterogeneous membrane containing AgCNS. It is used for the determination of the activity of SCN^- in suspensions of submitochondrial particles and chromatophores from *Rhodospirillum rubrum* in order to measure the membrane potential of these particles [200], as well as for other applications.

6.2 The sulphide ion-selective electrode

The properties of this electrode approach those of an ideal specific electrode. It has an anionic response almost exclusively for sulphide ions in a broad activity interval (up to 10^{-19} molar concentration in hydrogen sulphide solutions). Only the cyanide ion interferes here at high concentrations. This electrode has a cation response for silver ions in a wide activity range (in Ag(I) complexes up to 10^{-20} M Ag^+ [325]), where only Hg^{2+} interferes [417].

Silver sulphide exists in two modifications, α-Ag_2S, the cubic form, which is an electronic conductor and is stable above $176\,^\circ$C, and monoclinic β-Ag_2S, an ionic conductor, which is stable at lower temperatures [316]. In this latter modification, Ag^+ is almost the only charge carrier [141, 325, 428]. The good conductivity and negligible solubility of the compact membrane make the Ag_2S ISE one of the most reliable sensors.

The standard potential of these ISEs, analogous to halide ISEs, depends on the activities of silver and sulphur in the membrane. When metallic silver is used as a contact soldered directly onto the membrane, the silver activity in the membrane equals one and the Ag_2S ISE has properties identical with a silver electrode of the first kind or with a silver sulphide electrode of the second kind, depending on the solution with which it is in contact [203]. The membrane that is in contact with electrolyte on both sides behaves similarly [106].

Electrodes in contact with an electronic conductor other than silver, such as graphite, behave differently. Consider the cell [203] (cf. [334])

$$\begin{array}{c|c|c|c|c} \text{C} & \text{Ag} & \text{Ag}^+(\text{H}_2\text{O}) & \text{Ag}_{2+\delta}\text{S} & \text{C} \\ 4' & 1 & 2 & 3 & 4 \end{array}, \tag{6.2.1}$$

where δ is defined as the excess silver in Ag_2S. The EMF of this cell is given by the difference between the electrochemical potentials of the electrons in chemically identical graphite contacts 4 and 4'. As silver sulphide has a certain electronic conductivity, contact equilibrium may be formed at phase boundary 3/4. Thus, using the usual approach in which equilibrium between phases in contact characterizes equality of the electrochemical potentials of components present in both phases, it follows that

$$\mu_{Ag}^{(3)} - \mu_{Ag}^0 = -FE, \tag{6.2.2}$$

where μ_{Ag}^0 is the standard chemical potential of metallic silver and $\mu_{Ag}^{(3)}$ is the chemical potential of silver in Ag_2S. This equation is valid when a diffusion potential is not formed in the Ag_2S phase (cf. section 3.4) and when phase 2 does not contain redox systems.

If the graphite contact in cell (6.2.1) is replaced by a silver contact, quantity δ corresponds to saturation of silver sulphide with silver and the silver activity in Ag_2S equals unity. Then $\mu_{Ag}^{(3)} = \mu_{Ag}^0$ and $E = 0$. The standard potential of the Ag_2S ISE is identical with the standard potential of the silver electrode, $E_{0,\text{ISE}} = E_{Ag^+/Ag}^0 = +0.799$ V versus SCE. On the other hand, when Ag_2S is in equilibrium with elementary sulphur (for example, when Ag_2S is prepared in an oxidizing medium [417]) then

$$\mu_{Ag}^{(3)} = 0.5\mu_{Ag_2S}^0 - 0.5\mu_S^0. \tag{6.2.3}$$

In this equation, the standard chemical potential $\mu_{Ag_2S}^0$ can be used because the deviation of δ from stoichiometry is much less than 1. According to the definition of the standard potential, $\mu_S^0 = 0$, is the standard Gibbs energy of the element in the standard state ΔG_f^0. Then

$$\mu_{Ag}^{(3)} = 0.5\mu_{Ag_2S}^0 = 0.5\Delta G_f^{0,298} = 39.16 \text{ kJ mol}^{-1}$$

and the EMF of cell (6.2.1) is 0.203 V. The resultant standard potential of the ISE is 1.002 V. The values of 0.799 V and 1.002 V represent the lower and upper limits of the standard potential of an Ag_2S ISE in contact with an 'inert' conductor such as carbon.

The Ag_2S ISEs used in practice are of three kinds: with a heterogeneous membrane containing Ag_2S precipitate (preferably precipitated from excess sulphide or by the action of H_2S [228]) in a matrix of silicon rubber [346] or a thermoplastic material [248], with a pressed pellet as a membrane [235, 325, 452] and with an Ag_2S single crystal membrane [417]. The condition of stoichiometric composition must be carefully maintained in the preparation. Sintered ISEs with silver selenide and telluride have also been proposed [156a,

258]. Veselý and coworkers [417] listed the preparation methods, basic properties and response of this electrode for S^{2-}, Hg^{2+} and CN^-, as well as earlier unsuccessful attempts to make such an electrode. A Selectrode using Ag_2S [328] and an ISE with contact made of conductive artificial resin [83] have also been described.

The Ag_2S ISE has Nernstian response $dE/d \log a_i = \pm 0.0296$ V in the sulphide concentration range 10^{-2} to 10^{-7} M and silver ions from 10^{-1} to 10^{-7} M if the solutions are prepared from pure salts, as a further concentration decrease is prevented by adsorption on the glass (see p. 76 and [87, 163]). After prolonged use, the limit of the Nernstian behaviour shifts to about 10^{-6} M [130] as a result of formation of mixed potentials on accumulation of metallic silver in the membrane surface. An analogous deterioration in the membrane function in the presence of iodine results from surface oxidation [23]. Cyanide interferes only at large concentrations; the equilibrium constant of the reaction

$$6CN^- + Ag_2S = S^{2-} + 2Ag(CN)_3^{2-}$$

is about 10^{-2} [417]. It has been recommended that this electrode be calibrated using a sulphide buffer with an antioxidant [76]. This electrode has also been modified for flow-injection analysis of sulphides and thiols [83].

The principal application of the silver sulphide ISE is in the determination of S^{2-} in water [26, 86, 230, 283, 284, 440] (for a review see [167]), in waste waters from the paper industry after conversion of polysulphides and sulphites [47, 136, 293, 352], and as an indicator electrode in various direct or indirect argentometric titrations. Concentrations of the order of several ppb of H_2S in the air can be determined by argentometric titration [93]. In titrations the Ag_2S ISE has certain advantages over a silver electrode, but the response time is somewhat longer [72]. Its response to Hg^{2+} ions, while not ideal, makes it useful as an indicator electrode in mercurimetric determination of thiosulphate in the presence of sulphides, in the simultaneous determination of S^{2-}, $S_2O_3^{2-}$ and phenylmercaptan and for the determination of polysulphides [176, 293, 294], as well as for determining S^{2-} by titration with Pb(II) [270, 370]. Mercaptans and thiophene derivatives can be identified after gas chromatographic separation and conversion to H_2S [206, 207]. For applications in organic analysis see [339a, 339b].

The Ag_2S ISE is useful as an indicator electrode in the argentometric titration of thiol groups in proteins [128, 135]. The automatic determination [8, 9] depends on measuring the activity of excess silver ions after addition of Ag^+ to the protein and formation of silver mercaptan with thiol groups. This method is suitable for clinical analysis of changes in the overall protein and the albumin/ globulin ratio in serum. Monitoring of the antibody-antigen precipitation reaction is based on the same method [10]. The immunoglobulin fraction of goat anti-serum for the albumin of human serum precipitates the antibody prior to determination of the remaining antibody in the supernatant liquid. The determination is carried out using an ISE either after dissolving the precipitate in dilute alkali hydroxide or by analysis of the supernatant. This procedure has also been automated [10]. The Ag_2S ISE also reacts to cystein [396] and has been used for studying o-acetylserine sulphydrylase [272].

6.3 Electrodes containing divalent metal chalcogenides

Because of their very low conductivity, sulphides and other chalco-genides of divalent metals (except for electrodes with nonstoichiometric copper salts) are not suitable for preparation of ISEs. In contrast, membranes of pressed or sintered mixtures of the chalcogenides of lead, cadmium or Cu(II) with Ag_2S [325] (or Ag_2Se, Ag_2Te, HgS, HgSe, HgTe [335]) are sufficiently conductive and can be used for analytical purposes. The conduction mechanism is not clear [250], as CdS, PbS and CuS are insulators with negligible conduc-tivity and Ag_2S is a solid electrolyte. The function of these mixed material electrodes is a result of the response of the electrode to the silver ion activity, which is affected by the sulphide activity, depending on the activity of the divalent metal in the solution [325]. It thus holds that

$$a_{Ag^+}^2 a_{S^{2-}} = P_{Ag_2S} \tag{6.3.1}$$

$$a_{M^{2+}} a_{S^{2-}} = P_{MS} \tag{6.3.2}$$

from which it follows that

$$a_{Ag^+} = (P_{Ag_2S} a_{M^{2+}}/P_{MS})^{1/2}. \tag{6.3.3}$$

The ISE potential is then given by the relationship

$$\begin{aligned} E_{ISE} &= \text{const} + (RT/F) \ln a_{Ag^+} \\ &= \text{const} + (RT/2F) \ln (P_{Ag_2S}/P_{MS}) + (RT/2F) \ln a_{M^{2+}}. \end{aligned} \tag{6.3.4}$$

The electrode is, therefore, selective for M^{2+} ions in solution assuming that (*a*) the solubility product MS is much larger than P_{Ag_2S}, (*b*) the solubility of MS is much less than the concentration of M^{2+} ions, (*c*) the exchange reaction between MS and Ag_2S is sufficiently fast, which is fulfilled only for certain sulphides.

Equation (6.3.4) is valid for all-solid-state electrodes only when the membrane has a silver contact. The cell [203]

$$C \ \left| \ M \ \right| \ M^{2+}(H_2O) \ \left| \ \begin{matrix} M_{1+\delta}S \\ Ag_{2+\delta}S \end{matrix} \right| \ C$$

can be analysed in the same way as in section 6.2. It should be pointed out that this procedure is not suitable for membrane materials with ionic conductivity alone.

In acid media the sensitivity for divalent metals is decreased because of formation of $H_2 S$; consequently electrodes with selenides are preferable in acid media [155]. In the absence of complexes the electrodes have a response for divalent ions with Nernstian slope in the concentration range 10^{-1} to 10^{-4} mol dm^{-3}.

In the preparation of sulphide mixtures, it is suitable to carry out the precipitation by adding heavy metal salts to excess alkali metal sulphide [144, 408].

Ion-selective electrode for Pb^{2+} The commonly used version of the Pb^{2+} ISE has a pressed or sintered PbS – $Ag_2 S$ mixture membrane [144, 325]. The sintered membrane has the character of a solid solution [181]. The dependence of the potential of ISEs with various PbS : $Ag_2 S$ ratios on the Pb^{2+} activity is depicted in fig. 6.2. An ISE containing a mixture of PbS and $Ag_2 S$ in a plastic matrix has been proposed [249, 250]. Several versions of this ISE (PbTe, PbSe, PbS in an $Ag_2 S$ mixture) were tested in buffers for Pb^{2+} [242]. Hirata and Higashiyama [154, 156] suggested a mixture of PbS, CuS and $Ag_2 S$ for a Pb^{2+} ISE. This electrode is characterized by high stability, but also reacts to redox systems in the test solution. Hg^{2+}, Ag^+ and Cu^{2+} interfere in the response of Pb^{2+} ISEs.

The most important application of this electrode is the titration of sulphate with lead(II) salts in various media [104, 145, 149, 166, 246, 326, 338, 343, 381, 414]. The precision of the determination increases in mixed solvents, for example 70% ethanol [119] or dioxan/water [340]. A special ISE with a membrane consisting of a PbS, $PbSO_4$, $Ag_2 S$ and CuS mixture has been constructed for sulphate determinations [320].

Cd^{2+} ion-selective electrodes The usual version of the Cd^{2+} ISE contains a membrane of a sintered or pressed mixture of CdS and $Ag_2 S$ [121, 325, 408]. Membranes from sintered $Ag_2 S$, CuS and CdS mixtures [157] have also been proposed, similarly as for Pb^{2+} ISEs. CdS precipitate in a polyethylene matrix [250] or a CdS–$Ag_2 S$ precipitate mixture in a silicone rubber matrix [153] can also be used for Cd^{2+} ISEs. Cd^{2+} ISEs can be calibrated using a metal diethylenetriamine buffer [66]. Similar substances interfere in the response of the Cd^{2+} ISE as for the Hg^{2+}, Ag^+ and Cu^{2+} electrodes.

The Cd^{2+} ISE is used primarily in complexometric titrations with EDTA and Cd^{2+} as an indicator [327, 386]; this approach is also the basis of stability constant determination [268, 268a].

Ion-selective electrodes for Cu²⁺ and Cu⁺ There are two basic kinds of Cu ISE, one with Cu(II) alone in the membrane, the other with a mixture of Cu(II) and Cu(I).

In the electrode with CuS alone, which has negligible conductivity, the precipitate ISE with a silicone rubber matrix has better properties than the electrode with a pressed pellet [314]. The ISE with a mixture of CuS and Ag_2S finds broad application [325]. If the membrane is prepared by pressing, the grains of these two compounds combine to form jalpaite, $Ag_{1.55}Cu_{0.45}S$ [180]. This substance is a mixed conductor with transport numbers of Ag^+, 0.69; Cu(I), 0.30; and electrons, 0.01, at 25 °C [175]. The sintered electrode also contains $Ag_{1.2}Cu_{0.8}S$ or $Ag_{0.93}Cu_{1.07}S$. Oxidation of these phases leads to considerable deterioration in the electrode function [180]. Good electrodes

Fig. 6.2. The dependence of the response of the Pb(II) ISE on the ratio PbS: Ag_2S. The membrane consists of PbS and varying contents of Ag_2S: 1 – 0; 2 – 0.05 mol%; 3 – 0.25 mol%; 4 – 0.5 mol%; 5 – 1 mol%; 6 – 10 mol%; 7 – 40 mol%; 8 – 75 mol%; 9 – 85 mol%; 10 –95 mol%. (After Ito *et al.* [181].)

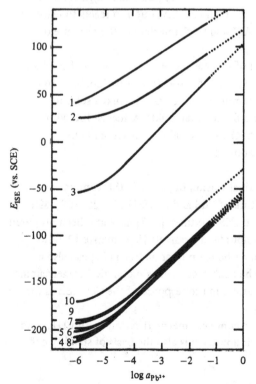

can be prepared from chalcocite (Cu_2S), either sintered [152] or as a single crystal [168], from $Cu_{2-x}S$ precipitate [301, 302] and from the $Cu_{2-x}Se$ single crystal [4, 5, 412]. The structures of the surfaces of various types of Cu ISE have been studied by powder x-ray diffractometry, scanning electron microscopy [142], Auger electron spectroscopy [261], ESCA [59] and XPS-photoelectron x-ray spectroscopy [90, 91, 314]. In the latter work, the formation of $CuSO_4$ caused by the action of oxidants results in the development of cracks [188]. The electrode can be regenerated by ascorbic acid (see also [251, 314, 391]).

A good deal of attention has been devoted to the effect of chlorides on the behaviour of Cu ISEs [221, 325, 441] (for a review see [129]). While in the absence of chloride the ISE potential depends linearly on the logarithm of the Cu^{2+} activity, with Nernstian slope of 0.0296 V (25 °C), at chloride concentrations greater than the Cu^{2+} concentration the slope changes to 0.0591 V and a marked response to the chloride ion activity appears. Westall *et al.* [441] state that contact of Cu^{2+} with the membrane leads to electrode oxidation,

$$Cu^{2+} + CuS \rightleftarrows 2Cu^+ + S.$$

In the presence of chloride the $CuCl_2^-$ complex is formed, shifting the equilibrium at the electrode surface to the right. At a sufficiently large Cl^- concentration, the ISE potential obeys the relationship

$$E_{ISE} = \text{const} + \frac{RT}{F} \ln\left(\frac{a_{Cu^{2+}}}{a_{Cl^-}^2 \beta_{CuCl_2^-}}\right), \tag{6.3.5}$$

where $\beta_{CuCl_2^-}$ is the overall stability constant of the $CuCl_2^-$ complex. Ag^+ and Hg^{2+} are the main interferents in the response of the Cu ISE.

The various versions of the Cu ISE also include flow-through electrodes [36, 389], porous flow-through electrodes [290] and flow-through microelectrodes [407].

The most important applications of Cu ISEs are in the direct determination of Cu^{2+} in water [169, 372, 410], complexometric titration of various metal ions using Cu^{2+} as an indicator [30, 143, 269, 385] and complexometric titrations of Cu^{2+} [409]. This ISE has also been used in the determination of the equilibrium activity of Cu^{2+} in various Cu^{2+} complexes in order to determine the stability constants (see [46, 285, 317, 318, 427, 445]), in the determination of the solubility of poorly soluble salts [122] and in the determination of the standard Gibbs transfer energies [58]. It can also be used in concentrated electrolytes [170].

Other systems A chalcogenide Zn ISE [454] and an Hg ISE with an HgS or HgSe membrane in an epoxide matrix [446] have been proposed.

6.4   The lanthanum trifluoride ion-selective electrode

The determination of fluoride ions has always been difficult, so the discovery of the lanthanum trifluoride electrode by Frant and Ross in 1966 was a great step forward. Until recently, this was the most important sensor in the ISE field, except for the glass electrode sensitive to hydrogen ions. The extraordinary specificity of this electrode made the greatest contribution to its usefulness. The only important interferent is the hydroxide ion.

Properties of LaF₃ crystals The rare earth fluorides crystallize in the hexagonal system (lanthanum trifluoride type lattice) or in the rhombic system (yttrium trifluoride type lattice). It should be mentioned that CeF_3, PrF_3, NdF_3 and SmF_3 also have LaF_3-type lattices. In this type of lattice, the La^{3+} ion is surrounded by five fluoride ions and the six next nearest neighbours are also fluoride ions. Consequently, the lattice consists of layers of LaF_2^+ with a fluoride layer on either side. In the equilibrium position, the fluoride ions lie in a shallow potential well (which is partly due to their radius of 0.133 nm; apart from the hydride ion, this is the smallest anion). Thus the fluoride ions are relatively mobile in the crystal lattice [14, 89, 120, 325, 332, 345, 451], so that LaF_3 has the properties of a solid electrolyte. Its conductivity is 3.6 to 2.9×10^{-7} S cm^{-1} at 25 °C [15]. Charge transfer is depicted by the scheme [89, 363, 404]:

$$LaF_3 + \text{molecular hole} \rightarrow LaF_2^+ + F^-.$$

A necessary condition for the function of the LaF_3 ISE is that the membrane consist of an LaF_3 single crystal, usually doped with europium [109, 415]. The importance of the presence of europium depends on the increased conductance of the membrane. At 25 °C the membrane doped with 0.5% wt. EuF_3 has a conductivity of 1.9 to 2.4×10^{-6} S cm^{-1} and with 1% wt. EuF_3, $2.8 - 4.3 \times 10^{-6}$ S cm^{-1} [15]. It could be expected that NdF_3 and SmF_3 would also be suitable, though too expensive, membrane materials [14, 325], while rare earths other than europium as well as Ca^{2+} [420] would be useful for doping. In contrast, other fluoride materials or LaF_3 precipitate in a polymeric matrix are not suitable.

At an ionic strength of $I = 0.08$ M, the solubility product of freshly precipitated LaF_3 [234] is $P_{LaF_3} = 10^{-17.9}$. However, compact LaF_3 [53] should, according to the detection limit of the LaF_3 ISE in fluoride ion solutions, have a much lower solubility product, approximately $10^{-24.5}$. This large difference between the solubility products is probably connected with the crystal surface energy, which is much higher for freshly precipitated polycrystalline material than for a single crystal. Another possible explanation of this difference assumes that, in contact with water, the compact form of LaF_3 dissolves very slowly and that the concentration of F^- at the membrane/solution boundary is given by the kinetics of dissolution and the rate of transport from the membrane, which is supported by the influence of stirring on E_{ISE} at low concentrations [376]. The low solution rate is caused by strong bonding of lanthanum ions in the crystal lattice [236] (see also [139, 255]).

Characteristics of the fluoride ion-selective electrode The LaF_3 ISE exhibits Nernstian response to the activity of fluoride ions in the concentration range 1 M to 10^{-6} M [31, 53, 325]. When the solution is buffered for fluoride ions using ZrO^{2+} and Th^{4+} salts, Baumann [28] has demonstrated that Nernstian response can be obtained down to a free fluoride ion concentration of 10^{-10} mol dm^{-3}. In solutions with an ionic strength of 2 mol dm^{-3}, the detection limit is 0.2 ppm, i.e. about 10^{-6} mol dm^{-3}, while in pure NaF solutions it may be as low as 10^{-7} mol dm^{-3}.

The LaF_3 ISE potential is affected by the presence of interfering ions in solution that affect the activity of the fluoride ions by formation of complexes, for example aluminium(III), ferric or beryllium(III) ions, or oxonium ions by formation of HF_2^- species in acid medium [27, 37, 256].

In contact with water, the reaction

$$LaF_3 + H_2O \rightarrow LaOF + 2HF$$

occurs at the LaF_3 surface [449]. The O^{2-} and OH^- ions have slightly greater radii than F^-, so that they can readily enter the crystal lattice and change the properties of LaF_3. In the presence of hydroxide ions, various solid solutions with composition $La(OH)_{1-x}F_{2+x}$ are formed [244, 307]. Consequently, the only important interferent is the hydroxide ion [109], as indicated in fig. 6.3. Frant and Ross [109] characterized the selectivity with respect to hydroxide ions by the coefficient $k_{F^-, OH^-}^{pot} = 0.1$; in actual fact, this value increases from very small values for 0.1 M F^- to a value greater than one in 10^{-5} M F^- [53]. It is also time-dependent, especially in alkaline solutions. Penetration of hydroxide ions into the membrane leads to the formation of a diffusion potential (see section 3.4) so that LaF_3 ISEs have certain properties similar to glass electrodes [43, 416]. Potentiometry, x-ray analysis, IR spectroscopy and tritium radiometry have demonstrated competitive adsorption of F^- and OH^- in the hydrophilic film on the LaF_3 crystal wall, affecting the ISE response [418].

The analytical properties of the fluoride ion-selective electrode At present, practically only ISEs constructed from a LaF_3 single crystal doped with EuF_3 and with internal metal contacts are used (see p. 65). A microelectrode for 2 μl of sample has also been constructed [85]. The standard type of LaF_3 ISE can be readily adjusted for measurements on nanolitre [425] and microlitre [426] samples (see fig. 6.4). Direct measurements can be reliably carried out in samples with concentrations of about 1 ppb, if the procedure recommended in [432] is maintained (for the determination of nanogram and subnanogram amounts of F^-, see [73]).

The response time of the LaF_3 ISE is less than half a second in solutions with concentrations greater than 10^{-3} M and less than three minutes in 10^{-6} M solutions [315, 376, 377].

Frant and Ross [108] recommended sample adjustment using TISAB buffer ('Total Ionic Strength Adjustment Buffer'), obtained by dissolving 57 ml glacial acetic acid, 58 g NaCl and 4 g 1,2-cyclohexanediaminetetraacetic acid (CDTA), adjustment of the solution pH with sodium hydroxide to 5 to 5.5 and dilution to 1 litre, all to maintain a constant ionic strength and pH between 5 and 5.5 and to complex ions such as Al^{3+} or Fe^{3+} that interfere in the determination. A detailed

Fig. 6.3. The effect of pH on the potential of the Orion Research lanthanum fluoride electrode in NaF solutions of various concentrations. The potential change with pH in the acidic region is caused by the formation of HF_2^-. (After Butler [53].)

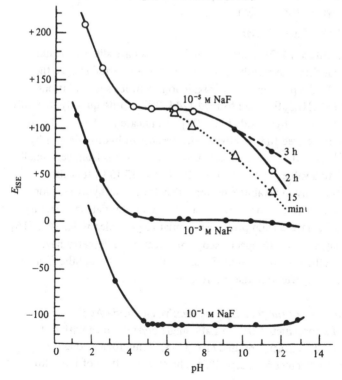

discussion of eleven different versions of the TISAB buffer [273] led to the conclusion that it is best to add 243 g of triammonium citrate to this buffer for determination of F^- in waters. To ensure complete decomposition of fluoride complexes of heavy metals, it is necessary to leave the solution for at least 20 minutes after sample mixing with the buffer (and preferably 24 hours). At very high Al^{3+} concentrations, aluminium must be separated by extraction of the complex with quinoline-8-ol into chloroform [279]. A buffer containing $HClO_4$, citric acid and triethanolamine [81] has been recommended for the determination of F^- in phosphates.

LaF$_3$ ISEs can also be used in organic and mixed solvents [116, 243, 304]. A marked increase in the sensitivity was observed in solutions containing dioxan [387].

Analytical determinations with the fluoride ion-selective electrode These are based either on direct potentiometry of fluorides [37, 84, 85, 88, 430] or on titration determinations of either fluorides or of other ions and also on titrations with fluoride ions as indicator. The advantages of potentiometry with an ISE over other analytical methods for determining fluorides were pointed out by Crosby *et al.* [67]. Further comparison studies [42, 56, 191, 433] came to the same conclusions, confirmed also by a study of 16 methods [365]. Fluoride ions are titrated either with La^{3+} (for concentrations greater than 1 mM) or Th^{4+} (in the concentration range 0.2-1 mM F^-) [13, 102, 103, 113, 233, 234]. Titration with fluoride ions can be used for the determination of Al^{3+} with formation of the AlF_4^- complex up to nanomolar concentrations, especially in ethanol–water mixtures [25] (see also [267, 384]). Precipitation titrations can also be used to determine La^{3+}, Th and UO_2^+ [241, 384] as well as Li^+ in

Fig. 6.4. A fluorideISE for measurement of nanolitre samples. (After Vogel and Brown [425].)

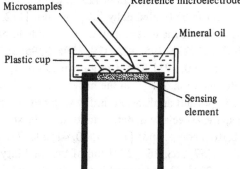

ethanolic solutions [24]. The fluoride electrode can also be used as an indicator electrode in the titration of Fe(III) using EDTA after addition of one drop of 10^{-3} M NaF to the test solution [342]. The micro- and semimicro-determination of AsO_4^{3-} is based on precipitation of La^{3+} and subsequent titration of excess La^{3+} with a fluoride solution, using a LaF_3 ISE for indication [337].

The sensitivity of determination of traces of F^- can be increased by extraction with tetraphenylantimony(V)dichloride in CCl_4 in the presence of CDTA and phosphate buffer. After back-extraction into an aqueous solution, the extractant is removed by shaking with CCl_4. This concentrated solution is analyzed using a fluoride ISE and down to 10^{-8} M F^- can be determined in the original solution [55*b*]. For determination of F^- after electrochemical generation of fluoride, see [88].

Applications of fluoride ion-selective electrodes The most important application of these electrodes is in the analysis of potable, waste and sea waters [67, 137, 167, 198, 230, 323, 330, 331, 429, 431–433] and of the atmosphere [33, 50, 100 148, 162, 229, 253, 271, 286, 309, 390, 448]. This electrode finds other important uses in phosphate analysis [48, 56, 79–81, 92, 289, 369, 402, 403] and in analysis of other rocks and minerals [38, 131, 177, 182, 287, 298, 406, 436] and soils [65, 159, 199, 224, 252, 254], either after dissolution in mineral acids or after fusing with $ZnO + Na_2CO_3$. In biomedical analysis (for a review, see [110]) a method has been developed for the determination of F^- in plasma by the standard addition method, where down to $2.5\,\mu g/l\,F^-$ can be determined [111] (see also [55*c*, 63, 179, 367, 411]). A method for determining F^- in urine is also useful [55, 55*c*, 62, 63, 179, 292, 400, 401], where fluoride concentrations are higher than in plasma. Another important application in this region is the determination of F^- after mineralization in bones [20, 303, 366, 453] and in tooth enamel [2, 35, 49, 105, 147, 223]. A study of F^- release from fluoride containing mouth rinses was also made [1, 173]. F^- is determined in plants after mineralization by burning in O_2 in a Schöninger flask [227]; using a semiautomatic method, up to 60 samples can be treated daily [50] (see also [172, 419]). Another important application field is the determination of fluoride in organic compounds [3, 107, 138, 146, 161, 232, 295, 322, 339, 339*a*, 339*b*, 362, 388] and in pharmaceuticals [52, 115, 190, 191]. The titration with fluoride ions has been used in the determination of Al^{3+} in the paper industry [284*a*].

Fluoride ISEs have found important applications in the study of equilibria in which fluoride ions participate, especially in determination of the stability constants of fluoride coordination compounds [18, 29, 40, 41, 53*a*, 71, 101, 114, 123, 133, 140, 220, 241*a*, 257, 265, 364, 377] and of the solubility products of poorly soluble fluorides [39, 324]. They are used in the kinetic study of reactions in which fluoride ions participate [16, 17, 377, 378, 399].

6.5 Glass electrodes

The glass electrode for pH measurements has long been a standard laboratory sensor and the subject of several monographs [74, 118, 349]. Consequently, only the basic theory will be given here, though electrodes sensitive to the alkali metal ions and silver will be considered in somewhat greater detail [95].

The dependence of the membrane potential of thin-walled glass bubbles on the acidity was known almost eighty years ago, but systematic investigation of the response to hydrogen ions and its dependence on the glass composition began in the 1920s (for a review of these works, see [178]). The most important advances were made by Hughes [164], and MacInnes and Dole [239, 240], who suggested an optimum glass composition of 22% wt. Na_2O, 6% wt. CaO and 72% wt. SiO_2. This glass is manufactured by Corning Glass Co. as Corning 015. Sokolov and Passynskii [373] independently developed a similar glass.

These glass electrodes are characterized by Nernstian response to hydrogen ions up to pH 11–12. Deviations at higher pH values depend on the kind and concentration of alkali metal cations present in the solution. To overcome this defect, a glass with a composition similar to that described has been proposed, in which sodium is replaced by lithium or potassium. It was found that lithium glass retains a negligible deviation from Nernstian response in 0.1 M KOH and only a slight deviation in 0.1 M NaOH. A glass containing 14.3% wt. Li_2O, 7% wt. BaO and 78.8% wt. SiO_2 [54] and a lithium glass containing 1–2 mol.% Cs_2O and several mol.% La_2O_3 or CeO_2 [297] were found to be best.

The 'alkaline error' of glass electrodes is basically caused by selectivity for alkali metal ions. This fact was originally expressed quantitatively by Nikolsky and Tolmacheva [278] and led Lengyel and Blum [226] to prepare a glass containing, among other components, aluminium and boron silicates. This glass exhibits selectivity for the alkali metal ions in a broad pH range. The sensitivity of the glass for alkali metal ions depends on the presence of oxides of trivalent metals in the glass. If their content is greater than 10 mol.%, glass electrodes are obtained that exhibit Nernstian response toward 10^{-3} M to 1 M Na^+ from pH 6 to 10. The selectivity for alkali metal ions was studied systematically by Eisenman and coworkers [99], who finally developed the NAS 11-18 electrode for determination of Na^+ [96, 178] and the NAS 27-06 electrode for determination of K^+ (the numbers indicate the mole percentage of Na and Al). Sodium ions can be determined in the presence of a large excess of potassium ions using a glass with k_{K^+, Na^+}^{pot} values of up to 250. Even greater selectivity has been described for glasses containing Si, Al, Li, B and Ga [379, 380].

The use of glasses selective for K^+, which should contain as much Na_2O or K_2O as possible, is limited by their durability, which decreases with increasing Na_2O content. Consequently, useful glass with k_{K^+, Na^+}^{pot} less than 0.1 cannot be obtained [178, 365a]. Table 6.1 lists several kinds of glass proposed by

Table 6.1. *Glass composition for glass electrodes suitable for determining* Li^+, Na^+, K^+ *and* Ag^+ [319].

Cation	Glass composition (in mol. %)	Selectivity	Remarks
Li^+	15% Li_2O–25% Al_2O_3–60% SiO_2	$k^{pot}_{Na,Li} \approx 3$ $k^{pot}_{K,Li} > 1000$	The best for Li^+ in the presence of H^+ and Na^+
Na^+	11% Na_2O–18% Al_2O_3–71% SiO_2	$k^{pot}_{K,Na} \approx 2800$, pH 11 $k^{pot}_{K,Na} \approx 300$, pH 7	Nernstian response up to $pNa^+ \approx 5$
Na^+	10.4% Li_2O–22.6% Al_2O_3–67% SiO_2	$k^{pot}_{K,Na} \approx 10^5$	Very selective towards Na^+, but a large drift
K^+	27% Na_2O–5% Al_2O_3–68% SiO_2	$k^{pot}_{Na,K} \approx 20$	Nernstian response up to $pK^+ < 4$
Ag^+	28.8% Na_2O–19.1% Al_2O_3–52.1% SiO_2	$k^{pot}_{H,Ag} \approx 10^5$	Very Ag^+-sensitive, but unstable
Ag^+	11% Na_2O–18% Al_2O_3–71% SiO_2	$k^{pot}_{Na,Ag} \approx 1000$	Less Ag^+-sensitive, but more reliable

Rechnitz [319] for the determination of Li^+, Na^+, K^+ and Ag^+ (cf. also [299]), together with their selectivity characteristics. Glass electrodes may also be selective for NH_4^+ that can be determined using a potassium electrode at low potassium ion concentrations [95].

So far, a sufficiently durable glass with good selectivity for divalent cations, especially Ca^{2+} has not been obtained [394]. Phosphate glass does exhibit such selectivity [395], but its technical development is in the initial stages.

Glass microelectrodes are discussed in section 4.2.

Initial attempts to explain the membrane potential of glass electrodes as a simple Nernst potential with diffusible hydrogen ions successfully explained the Nernstian response to these ions, but could not explain the selectivity for alkali metal ions. In addition, they provided an unrealistic explanation for the mechanism of formation of the membrane potential. Hydrogen ions do not pass through the membrane at all, as was clearly demonstrated by Schwabe and Dahms [350] using a coulometric method where the glass electrode acted as a cathode in a solution containing tritium. Even after about 20 hours electrolysis, tritium did not appear on the inner side of the glass membrane. Establishment of a membrane potential requires, of course, at least small conductivity of the membrane, provided by monovalent ions in the membrane [189].

The membrane potential formed at glass membranes is a result of ion exchange between the solution and membrane, as follows clearly from the works of Schiller [344], Horowitz [160], Dole [75] and Nikolsky [274], whose results are reviewed in [277]. The electrical potential difference at the solution/ glass membrane phase boundary is a function of the cation activity ratio, for example of hydrogen ions, in the solution and in the membrane. The cation activity in the membrane depends on the ion-exchange equilibrium (3.2.13), so that the simple expression (3.1.4) is valid only if the equilibrium of the exchange reaction is shifted so that one kind of cation predominates in the membrane and is also present in solution. This occurs, for example, with glass electrodes made of Corning 015 glass if the solution pH is less than 12. If however the membrane simultaneously contains both kinds of ions present in the solution, a diffusion potential is formed in the membrane for equilibrium at the solution/membrane boundary. Equation (3.4.1) must be used to characterize this situation.

A glass membrane in an electrolyte solution cannot be taken to be a homogeneous system in the direction perpendicular to the surface. When the membrane is in contact with the solution, water molecules can enter it and form a 5–100 nm thick hydrated layer [319]. The formation of this hydrated layer is actually a condition for good functioning of the glass electrode. The basic characteristics of the glass structure probably do not change in the hydrated layer, but the cation mobility increases considerably compared with the compact membrane interior (the diffusion coefficient of monovalent cations in hydrated glass is about 5×10^{-11} cm^2 s^{-1} and in compact glass about a thousand times smaller [351]).

Transport of cations in glass is controlled by the defect mechanism where the defects are cations in interstitial positions [78]. The diffusion coefficient is given by the product

$$D = g\langle\lambda^2\rangle f, \tag{6.5.1}$$

where g is a geometric factor, λ is the distance of cation jump and f is the jump frequency defined by the relationship

$$f = PN, \tag{6.5.2}$$

where P is a value characterizing the jump probability and N is the number of cations in interstitial positions. This quantity depends on the Gibbs energy of formation of interstitial cation-hole pairs, ΔG_f, according to the relationship

$$N = p^{1/2} \exp\left(-\Delta G_f/2RT\right), \tag{6.5.3}$$

where p is the number of interstitial sites for a single cation. Thus, it follows for P that

$$P = \nu \exp\left(-\Delta G_m/RT\right), \tag{6.5.4}$$

where ν is the vibrational frequency of the diffusing cation ($\nu \approx 10^{13}$ s^{-1}) and ΔG_m is the Gibbs energy required for transfer of an ion from one interstitial site to another (i.e. a different mechanism from the Frenkel mechanism described on p. 131). The effect of structural factors, such as the cation radius and the effective O^{2-} radius in the glass, is included in ΔG_f which, according to the Madelung equation, is inversly proportional to the sum of the cation radius and the effective radius of the oxygen anion. This value is thus greater (and the ion mobility is smaller) for smaller cation and effective oxygen ion radii, because then their approach is closest.

The theory of glass membranes with two layers characterized by different ion exchange constants and different mobilities of the exchanged ions was developed by Conti and Eisenman [61]. Assuming a stationary state they obtained (3.4.1) or (3.4.4) expanded by an expression that is a function of the cation concentration at the boundary between the two membrane phases and their mobilities. This term can be considered constant.

Eisenman [98] verified his relationship for the selectivity coefficient (3.4.1) using a glass electrode selective for both potassium and sodium ions. First he found

$$k_{Na^+, K^+}^{pot} = K_{exch}(U_{K^+}/U_{Na^+})$$

from the dependence of the potential of the glass electrode on the ratio of the concentrations of the two ions. He then determined the ratio U_{K^+}/U_{Na^+} from study of the diffusion of the labelled form of one kind of ion into the electrode when the hydrated layer is completely saturated by the other ion (by prolonged storage in a solution of the carbonate of this ion). He finally determined the ion exchange constant K_{exch} using diffusion of labelled K^+ into a membrane saturated with Na^+ at various values of the K^+/Na^+ ratio in solution. The sodium ion is roughly ten times more mobile in hydrated glass than the potassium ion. The calculated and experimental values of the selectivity coefficient were in quite good agreement.

The selectivity of glass for alkali metal ions is connected with the presence of oxides of trivalent metals in the glass structure. Zachariasen [450] states that silicate glass has a random cross-linking, where each silicon atom lies in the centre of a tetrahedron formed of oxygen atoms (see planar scheme (6.5.5)).

$$\begin{array}{cc} O & O \\ | & | \\ -O-Si-O-Si-O- \\ | & | \\ O & O \end{array} \qquad (6.5.5)$$

The arrangement of the tetrahedra is not, however, as regular as in a crystal. The oxygen atoms connecting silicon atoms are termed 'bridging' oxygens. If the glass contains the monovalent cation M^+, a 'non-bridging' slightly charged oxygen is formed; in the presence of alkaline earth ions, two non-bridging oxygens are formed. Silicon can be replaced in the lattice by aluminium, but the excess negative charge of the group

$$\begin{array}{c} O \\ | \;^{(-)} \\ -O-Al-O- \\ | \\ O \end{array} \qquad (6.5.6)$$

must be neutralized, for example by an alkali metal cation.

The selectivity constant for ion K^+ with respect to ion J^+ depends on the exchange constant K_{exch}. This constant is given by the ratio of the stability constants of the associates of the given ions with the ion exchanger active site X:

$$K_{exch} = K_{JX}/K_{KX}. \qquad (6.5.7)$$

This problem is considerably simplified if we assume that the active sites are a single group, although in actual fact 6 to 8 AlO_4^- groups bond sodium and 8 groups bond potassium. Williams states [300, 443] that a relationship can be found between the associate stability and the ratio of the Pauling radii of the cation and the bonding anion in the following manner (also used by Eisenman [94, 97]).

Association of the solvated ion KS with the oppositely charged ion X, i.e.

$$KS + X \rightleftarrows KX + S \qquad (6.5.8)$$

is accompanied by a change in the Gibbs energy ΔG_K. This process may be separated into two steps by considering the equilibrium with gaseous ion K(g),

$$K(g) + S \rightleftarrows KS \text{ (Gibbs energy change } \Delta G_1)$$
$$K(g) + X \rightleftarrows KX \text{ (Gibbs energy change } \Delta G_2).$$

(6.5.9)

Apparently,

$$\Delta G_K = \Delta G_2 - \Delta G_1 \qquad (6.5.10)$$

and a complex is formed if $\Delta G_K < 0$. ΔG_1 can be expressed by the modified Born equation (see [214], chapter 1),

$$\Delta G_1 = -A/(r_K + 0.085) = -A/r_{eff}, \qquad (6.5.11)$$

where r_K is the Pauling radius of the cation in nm and A is a function of the cation charge and of the dielectric constant of the medium. It can be written for ΔG_2 that

$$\Delta G_2 = -B/(r_K + r_X), \qquad (6.5.12)$$

where r_X is the Pauling radius of X (must be greater than 0.085 nm) and B is directly proportional to the absolute value of the product of the charge numbers of K and X. Fig. 6.5 [444] depicts the dependence of ΔG_2 and ΔG_1 on the reciprocal effective radius of the ion r_{eff} and thus illustrates the conditions for associate formation.

It can be seen that the stability of the associate decreases with increasing radius K (curve 1), while for curve 2 the stability first increases and then decreases with increasing radius K. Apparently, in the presence of a large electric field of X (small r_X or large charge number of X), the stability (and thus also the selectivity for various ions K with respect to the given ion) depends linearly on the radius of this ion r_K, i.e. the selectivity for the lithium ion would be the greatest of that for all cations. This is also true for classical sodium–calcium glasses, where the non-bridging oxygens represent sites with strong electric fields (the selectivity for alkali metal ions is considerably overshadowed by the selectivity for H^+). However, for alumino-silicate glasses the field strength of the active sites is weakened and various selectivity orders may be observed for alkali metal ions. In this way, a qualitative explanation may be obtained for the increased selectivity of glass with a certain composition for K^+ with respect to Na^+ and Rb^+. Eisenman's attempt [97] to express these considerations quantitatively as a larger number of selectivity orders, assuming changes in the electric field intensity of the active sites, is too speculative in character.

Nikolsky and coworkers [275–277] developed a theory of the electric potential at the membrane/electrolyte phase boundary for exchange between the cations in solutions and variously active sites in the membrane. They successfully explained, among other things, the origin of the inflection point on the dependence of the glass electrode potential on the pH for some lithium–silicon glasses containing a small amount of Al_2O_3, Ga_2O_3 or B_2O_3. However, this theory does not consider the diffusion potential in the glass.

In conclusion, mention should be made of glass electrodes exhibiting mixed, ionic and electronic conductances. A chalcogenide glass (28% Ge, 60% Se, 12% Sb) electrode doped with Fe (max. 2%) has Nernstian response to Fe^{3+} and Cu^{2+} and also a certain response to nitrates [11, 19, 185]. The titration of sulphates with Ba^{2+} depends on the sensitivity of this ISE for Fe(III) ions present in the solution as an indicator [183]. This electrode was used for analysis of water [184]. Another electrode of chalcogenide glass [187] containing $Cu_6 As_4 S_9$ is sensitive for Cu^{2+} and has been used for the determination of Cu^{2+} in sea water at concentrations down to 1 ppb [186]. Other electrode types using chalcogenide glasses [291] exhibit selectivity for Cu^{2+} (membrane composition, $Se_{40-48}Cu_{21-35}\text{-}As_{25-30}$) or for Pb^{2+} (membrane composition, $Pb_{40-53}Se_{29-37}As_{18-23}$).

Fig. 6.5. The Gibbs energy for the formation of a hydrated cation from a gaseous cation (ΔG_1) and for the formation of three complexes (ΔG_2, curves 1,2,3) from gaseous cation and ligands of various types. Curve 1: a complex with the hydrated cation will not be formed; Curve 2: the ligand is an anion with a large charge or a small size; Curve 3: the ligand is an anion with a small charge or a large size. The real values of r_{eff} correspond to the region between the dashed straight lines. (After Williams [444].)

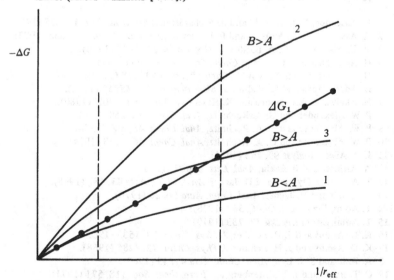

6.6 Various other systems

For more than the last half century experiments have been carried out to obtain electrodes with solid membranes containing inorganic insoluble compounds which the authors expect (consciously or not) to have ion-exchange properties (for a review, see [57]). These substances, for example insoluble cyanoferrates(II), phosphomolybdates, calcium fluoride, etc., are not, however, promising materials and probably only generate problems for the editors of specialized journals.

Electrodes with membranes containing the salts of organic radical ions [354–361, 435], for example the radical ion derived from 7,7,8,8-tetracyanoquino-dimethane,

or from N-ethylbenzothiazole-2,2-azine,

are more interesting systems. These ISEs are sensitive for the anions of the corresponding salts, especially the perchlorates.

References for Chapter 6

1 R. Aasenden, F. Brudevold and B. Richardson, *Arch. Oral. Biol.* **13**, 625 (1968).

2 R. Aasenden, E. C. Moreno and F. Brudevold, *Arch. Oral. Biol.* **17**, 355 (1972).

3 M. E. Aberlin and C. A. Bunton, *J. Org. Chem.* **35**, 1825 (1970).

4 H. Adametzová and J. Grégr, *Chem. Průmysl* **21**, 506 (1971).

5 H. Adametzová and M. Semler, *Chem. Průmysl* **23**, 418 (1973).

6 H. Adametzová and R. Vaďura, *J. Electrochem. Soc.* **55**, 53 (1974).

7 H. Akaiwa, H. Kawamoto and K. Hasekawa, *Talanta* **27**, 909, (1980).

8 P. W. Alexander and G. A. Rechnitz, *Anal. Chem.* **46**, 250 (1974).

9 P. W. Alexander and G. A. Rechnitz, *Anal. Chem.* **46**, 860 (1974).

10 P. W. Alexander and G. A. Rechnitz, *Anal. Chem.* **46**, 1253 (1974).

11 J. D. Allen, *Analyst* **99**, 765 (1974).

12 A. Altinata and B. Pekin, *Anal. Lett.* **6**, 667 (1973).

13 T. Anfält, D. Dyrssen and D. Jagner, *Anal. Chim. Acta* **43**, 487 (1968).

14 T. Anfält and D. Jagner, *Anal. Chim. Acta* **50**, 23 (1970).

14a T. Aomi, *Denki Kagaku* **46**, 567 (1978).

15 T. Aomi, *Denki Kagaku* **47**, 733 (1979).

16 K. D. Asmus and J. H. Fendler, *J. Phys. Chem.* **73**, 1533 (1969).

17 K. D. Asmus and J. H. Fendler, *J. Phys. Chem.* **72**, 4285 (1968).

18 A. Aziz and S. J. Lyle, *Anal. Chim. Acta* **47**, 49 (1969).

19 C. T. Baker and I. Trachtenberg, *J. Electrochem. Soc.* **118**, 571 (1971).

20 G. Bang, T. Kristoffersen and K. Meyer, *Acta Path. Microbiol. Scand.* A78, 49 (1970).
21 M. Bartušek, J. Šenkýr, J. Janošová and M. Polášek, Potentiometric determination of halide ions in solution, in *Ion-Selective Electrodes* (ed. E. Pungor), Symposium 1972, Akadémiai Kiadó, Budapest (1973), p. 173.
22 P. L. Bailey, J. Wilson, S. Karpel and M. Riley, Application of chloride electrodes based on mercurous chloride/mercuric sulphide, in *Ion-Selective Electrodes* (ed. E. Pungor and I. Buzás), Conference 1977, Akadémiai Kiadó, Budapest (1978), p. 201.
23 F. G. K. Baucke, *Z. Anal. Chem.* 282, 105 (1976).
24 E. W. Baumann, *Anal. Chem.* 40, 1731 (1968).
25 E. W. Baumann, *Anal. Chem.* 42, 110 (1970).
26 E. W. Baumann, *Anal. Chem.* 46, 1345 (1974).
27 E. W. Baumann, *Anal. Chim. Acta* 42, 127 (1968).
28 E. W. Baumann, *Anal. Chim. Acta* 54, 189 (1971).
29 E. W. Baumann, *J. Inorg. Nucl. Chem.* 31, 3155 (1969).
30 E. W. Baumann and R. M. Wallace, *Anal. Chem.* 41, 2072 (1969).
31 W. E. Bazzelle, *Anal. Chim. Acta* 54, 29 (1971).
32 C. G. Beguin and Ch. Coulombeau, *Anal. Chim. Acta* 90, 237 (1977).
33 J. Biheller and W. Resch, *Staub* 31, 9 (1971).
34 J. M. Birkenland, *Caries Res.* 7, 11 (1973).
35 T. J. Bitner and S. H. Y. Wei, *J. Dent. Res.* 52, 157 (1973).
36 W. J. Blaedel and D. E. Dinwiddie, *Anal. Chem.* 47, 1070 (1975).
37 R. Bock and S. Strecker, *Z. Anal. Chem.* 235, 322 (1968).
38 J. B. Bodkin, *Analyst* 102, 409 (1977).
39 A. M. Bond and G. Heftner, *Inorg. Chem.* 9, 1021 (1970).
40 A. M. Bond and G. Heftner, *J. Inorg. Nucl. Chem.* 33, 429 (1971).
41 A. M. Bond and T. A. O'Donell, *J. Electroanal. Chem.* 26, 137 (1970).
42 M. and B. Boniface, F. Erb and N. Hanquez, *Analysis* 2, 263 (1975).
43 M. J. Brand and G. A. Rechnitz, *Anal. Chem.* 42, 478 (1970).
44 T. Braun, C. Ruiz de Pardo and E. C. Salazar, *Radiochem. Radioanal. Lett.* 3, 419 (1970).
45 P. T. Bray, G. C. F. Clark, G. J. Moody and J. D. R. Thomas, *A Perspective of Sodium and Chloride Ion-Selective Electrode Sweat Tests for Screening in Cystic Fibrosis*, University of Wales, Cardiff (1975).
46 W. T. Bresnahan, C. Grant and J. H. Weber, *Anal. Chem.* 50, 1675 (1978).
47 G. Brunow, T. Ilus and G. E. Miksche, *Acta Chem. Scand.* 26, 1117 (1972).
48 L. G. Bruton, *Anal. Chem.* 43, 579 (1971).
49 C. Bruun, *Scand. J. Dent. Res.* 81, 92 (1973).
50 M. Buck and G. Reusman, *Fluoride* 4, 5 (1971).
51 R. P. Buck and V. R. Shepard, *Anal. Chem.* 46, 2097 (1974).
52 E. J. Bushee, D. K. Grissom and D. R. Smith, *J. Dent. Child.* 38, 279 (1971).
53 J. N. Butler, Thermodynamic studies, Chapter 5 of *Ion-Selective Electrodes* (ed. R. A. Durst), NBS Special Publication No. 317, Washington (1969).
53a J. Čadek, J. Veselý and Z. Šulcek, *Coll. Czech. Chem. Comm.* 36, 3377 (1971).
54 H. H. Cary and W. P. Baxter, US Patent 2462843 (1945).
55 A. A. Cernik, J. A. Cooke and R. J. Hall, *Nature* 227, 1260 (1970).
55a G. A. Cherian and G. J. Hill, *Clin. Chem.* 17, 642 (1971).
55b H. Chermette, C. Martelet, D. Sandino, M. Benmalek and J. Tousset, *Anal. Chim. Acta* 59, 373 (1972).
55c K. Chiba, K. Tsunoda, H. Haraguchi and K. Fuwa, *Anal. Chem.* 52, 1582 (1980).

56 R. L. Clements, G. A. Sergeant and P. J. Webb, *Analyst* **96**, 51 (1971).
57 C. J. Coetzee, *Ion-Sel. El. Rev.* **3**, 105 (1981).
58 J. F. Coetzee and W. K. Istone, *Anal. Chem.* **52**, 53 (1980).
59 J. F. Coetzee, W. K. Istone and M. Carvalho, *Anal. Chem.* **52**, 2353 (1980).
60 C. Collombel, J. P. Durand, J. Bureau and J. Cotte, *J. Eur. Toxicol.* **3**, 291 (1970).
61 F. Conti and G. Eisenman, *Biophys. J.* **5**, 247 (1965).
62 D. C. Cowell, *Ann. Clin. Biochem.* **14**, 269 (1977).
63 D. C. Cowell, *Med. Lab. Sci.* **35**, 265 (1978).
64 G. S. Craven and M. C. Griffith, *Austr. J. Dairy Technol.* **32**, 75 (1977).
65 G. L. Crenshaw and F. N. Ward, *Bull. U.S. Geol. Survey* No. 1408, 77 (1975).
66 M. Cromer-Morin and J. P. Scharff, *C. R. Acad. Sc. Paris* C **283**, 621 (1976).
67 N. T. Crosby, A. L. Dennis and J. G. Stevens, *Analyst* **93**, 643 (1968).
68 B. Csakvari and K. Meszaros, *Hung. Sci. Instr.* **11**, 9 (1968).
69 A. L. Cummings and K. P. Anderson, *Anal. Chem.* **47**, 2310 (1975).
70 H. J. Degenhart, G. Abelin, B. Bevaart and J. Baks, *Clin. Chim. Acta* **38**, 217 (1972).
71 J. DeMoura, D. LeTourneau and A. C. Wiese, *Arch. Biochem. Biophys.* **258** (1969).
72 A. Dencks and R. Neeb, *Z. Anal. Chem.* **285**, 233 (1977).
73 D. Deutsch and S. Zarini, *Anal. Chem.* **52**, 1167 (1980).
74 M. Dole, *The Glass Electrode*, John Wiley and Sons, New York (1941).
75 M. Dole, *J. Chem. Phys.* **2**, 862 (1934).
76 E. L. Donaldson and D. C. McMullan, *Anal. Lett.* A **11**, 39 (1978).
77 W. A. B. Donners and D. A. Wooys, *J. Electroanal. Chem.* **52**, 277 (1974).
78 R. H. Doremus, Diffusion potentials in glass, Chapter 4 of *Glass Electrodes for Hydrogen and Other Cations* (ed. G. Eisenman), M. Dekker, New York (1969).
79 E. J. Duff and J. L. Stuart, *Anal. Chim. Acta* **52**, 155 (1970).
80 E. J. Duff and J. L. Stuart, *Anal. Chim. Acta* **57**, 233 (1971).
81 E. J. Duff and J. L. Stuart, *Analyst* **100**, 739 (1975).
82 E. J. Duff and J. L. Stuart, *Talanta* **22**, 901 (1975).
83 E. J. Duffield, G. J. Moody and J. D. Thomas, *Anal. Proc. (London)* **17**, 533 (1980).
84 R. A. Durst, *Anal. Chem.* **40**, 931 (1968).
85 R. A. Durst, *Anal. Chem.* **41**, 2089 (1969).
86 R. A. Durst, Analytical techniques and applications of ion-selective electrodes, Chapter 11 of *Ion-Selective Electrodes* (ed. R. A. Durst), NBS Special Publication No. 317, Washington (1969).
87 R. A. Durst and B. T. Duhart, *Anal. Chem.* **42**, 1002 (1970).
88 R. A. Durst and J. W. Ross, *Anal. Chem.* **40**, 1343 (1968).
89 R. A. Durst and J. K. Taylor, *Anal. Chem.* **39**, 1483 (1967).
90 M. F. Ebel, W. Gröger, L. Pólos, K. Tóth and E. Pungor, *Hung. Sci. Instr.* **49**, 41 (1980).
91 M. F. Ebel, K. Tóth, L. Pólos and E. Pungor, *Surf. Interface Anal.* **2**, 197 (1980).
92 C. R. Edmond, *Anal. Chem.* **41**, 1327 (1969).
93 D. L. Ehman, *Anal. Chem.* **48**, 918 (1976).
94 G. Eisenman, *Biophys. J.* **2**, (2), Part 2, Suppl., 259 (1962).
95 G. Eisenman, Particular properties of cation-selective glass electrodes containing Al_2O_3, Chapter 4 of *Glass Electrodes for Hydrogen and Other Cations* (ed. G. Eisenman), M. Dekker, New York (1969).
96 G. Eisenman, The electrochemistry of cation-sensitive glass electrode, in *Advances in Analytical Chemistry and Instrumentation* (ed. C. W. Reilley), Vol. 4, J. Willey and Sons, Inc., New York (1965), p. 213.

97 G. Eisenman, Physical basis of the ionic specificity of the glass electrode, Chapter 7 of *Glass Electrodes for Hydrogen and Other Cations* (ed. G. Eisenman), M. Dekker, New York (1969).

98 G. Eisenman, The origin of the glass-electrode potential, Chapter 5 of *Glass Electrodes for Hydrogen and Other Cations* (ed. G. Eisenman), M. Dekker, New York (1969).

99 G. Eisenman, D. O. Rudin and J. U. Casby, *Science* **126**, 871 (1957).

100 L. A. Elfers and C. E. Decker, *Anal. Chem.* **40**, 1658 (1968).

101 B. Elquist, *J. Inorg. Nucl. Chem.* **32**, 937 (1970).

102 T. Eriksson, *Anal. Chim. Acta* **58**, 437 (1972).

103 T. Eriksson and G. Johansson, *Anal. Chim. Acta* **52**, 465 (1970).

104 W. Eysenbach, B. Sattkus and G. Heller, *Z. Anal. Chem.* **277**, 183 (1975).

105 F. T. Feagin, B. G. Jeansonne and A. McCaghken, *J. Dent. Res.* **51**, 1457 (1972).

106 K. S. Fletcher III and R. F. Mannion, *Anal. Chem.* **42**, 285 (1970).

107 H. J. Francis, J. H. Deomarine and D. D. Persing, *Microchem. J.* **14**, 580 (1969).

108 M. S. Frant and J. W. Ross, *Anal. Chem.* **40**, 1169 (1968).

109 M. S. Frant and J. W. Ross, *Science* **154**, 1553 (1966).

110 C. Fuchs, Solid-state ion-selective electrodes in clinical chemistry, Chapter 3 of *Medical and Biological Applications of Electrochemical Devices* (ed. J. Koryta), John Wiley and Sons, Chichester (1980).

111 C. Fuchs, D. Dorn, C. A. Fuchs, H. V. Henning, C. McIntosh, F. Scheler and M. Stennert, *Clin. Chim. Acta* **60**, 157 (1975).

112 C. Fuchs, D. Dorn and C. McIntosh, *Z. Anal. Chem.* **279**, 150 (1976).

113 F. F. Gaál, L. S. Jovanović and V. D. Canić, *Z. Anal. Chem.* **282**, 439 (1976).

114 H. Gamsjäger, P. Schindler and B. Kleinert, *Chimia* **23**, 229 (1969).

115 P. Farnier, J. Krembel, R. M. Frank and A. Bloch, *Rev. Fr. Odonto-Stomatol.* **18**, 613 (1971).

116 S. A. Gava, N. S. Poluektov and G. N. Koroleva, *Zh. Anal. Khim.* **33**, 506 (1978).

117 M. Geissler, *Z. Anal. Chem.* **302**, 188 (1980).

118 *Glass Electrodes for Hydrogen and Other Cations, Principles and Practice* (ed. G. Eisenman), M. Dekker, New York (1967).

119 J. O. Goertzen and J. D. Oster, *Proc. Soil. Sci. Soc. Amer.* **36**, 691 (1972).

120 M. Goldman and L. Shen, *Physical Rev.* **144**, 321 (1966).

121 A. V. Gordievskii, V. S. Sherman, A. Yu. Syrchenkov, N. T. Savvin and A. F. Zhukov, *Zh. Anal. Khim.* **27**, 772 (1972).

122 S. R. Grabler and S. K. Suri, *J. Inorg. Nucl. Chem.* **42**, 51 (1980).

123 S. L. Grassino and D. N. Hume, *J. Inorg. Nucl. Chem.* **33**, 421 (1971).

124 A. L. Grekovich, E. A. Materova and V. E. Jurinskaya, *Zh. Anal. Khim.* **27**, 1218 (1972).

125 T. B. Grimley, *Proc. Roy. Soc.* (A) **201**, 40 (1950).

126 O. Grof, *Vod. hospod.* **B22**, 41 (1972).

127 P. Grøn, H. G. McCann and F. Brudervold, *Arch. Oral Biol.* **13**, 203 (1968).

128 L. C. Gruen and B. S. Harrap, *Anal. Biochem.* **42**, 377 (1971).

129 J. Gulens, *Ion-Sel. El. Rev.* **2**, 117 (1980).

130 J. Gulens and B. Ikeda, *Anal. Chem.* **50**, 782 (1978).

131 J.-L-Guth and R. Wey, *Bull. Soc. Fr. Minér. Cristall.* **92**, 105 (1969).

132 R. Gyenge, E. Körös and K. Tóth, Calibration and application of the bromide ion-selective electrode in oscillating reactions in *Ion-Selective Electrodes* (ed. E. Pungor and I. Buzás), Symposium 1976, Akadémiai Kiadó, Budapest (1977), p. 115.

133 F. M. Hall and S. J. Slater, *Aust. J. Chem.* **21**, 1663 (1968).

134 M. N. Hall and A. A. Pilla, *J. Electrochem. Soc.* **118**, 72 (1972).

135 B. S. Harrap and L. C. Gruen, *Anal. Biochem.* **42**, 398 (1971).
136 W. C. Harris, E. P. Crowell and D. H. McMahon, *Tappi* **57**, 102 (1974).
137 J. E. Harwood, *Water Res.* **3**, 273 (1969).
138 S. S. M. Hassan, *Mikrochim. Acta* 889 (1974).
139 R. C. Hawkings, L. P. V. Corriveau, S. A. Kushnerik and P. Y. Wong, *Anal. Chim. Acta* **102**, 61 (1978).
140 G. Hefter, *J. Electroanal. Chem.* **39**, 345 (1972).
141 M. H. Hebb, *J. Chem. Phys.* **20**, 185 (1952).
142 G. J. M. Heijne and W. E. Van der Linden, *Anal. Chim. Acta* **93**, 99 (1977).
143 G. J. M. Heijne, W. E. Van der Linden and G. den Boef, *Anal. Chim. Acta* **98**, 221 (1978).
144 G. J. M. Heijne, W. E. Van der Linden and G. den Boef, *Anal. Chim. Acta* **100**, 193 (1978).
145 R. N. Heistand and C. T. Blake, *Microchim. Acta* 212 (1972).
146 L. Helešič, *Coll. Czech. Chem. Comm.* **37**, 1514 (1972).
147 E. Hepp and E. J. Duff, *Caries Res.* **10**, 234 (1976).
148 P. Hermann and W. Rode, *Staub - Reinhalt. Luft* **35**, 298 (1975).
149 J. E. Hicks, J. E. Eleenor and H. R. Smith, *Anal. Chim. Acta* **68**, 480 (1974).
150 K. Hiiro, T. Taneka, A. Kawahara and Y. Kono, *Bunseki Kagaku* **22**, 1072 (1973).
151 B. W. Hipp and G. W. Langdale, *Commun. Soil. Sci. Plant Anal.* **2**, 237 (1971).
152 H. Hirata and K. Date, *Anal. Chim. Acta* **51**, 209 (1970).
153 H. Hirata and K. Date, *Bull. Chem. Soc. Japan* **46**, 1468 (1973).
154 H. Hirata and K. Higashiyama, *Anal. Chim. Acta* **54**, 415 (1971).
155 H. Hirata and K. Higashiyama, *Anal. Chim. Acta* **57**, 476 (1971).
156 H. Hirata and K. Higashiyama, *Bull. Chem. Soc. Japan* **44**, 2420 (1971).
156a H. Hirata and K. Higashiyama, *Talanta* **19**, 391 (1972).
157 H. Hirata and K. Higashiyama, *Z. Anal. Chem.* **257**, 104 (1971).
158 S. Honda, K. Sudo, K. Karebi and K. Tariura, *Anal. Chim. Acta* **77**, 274 (1975).
159 D. M. Hopkins, *J. Res. U.S. Geol. Surv.* **5**, 589 (1977).
160 K. Horowitz, *Z. Physik. Chem.* **115**, 424 (1925).
161 K. Hozumi and N. Akimoto, *Japan Anal.* **20**, 467 (1971).
162 A. Hrabeczypál, K. Tóth, E. Pungor and F. Vallo, *Anal. Chim. Acta* **77**, 278 (1975).
163 T. M. Hseu and G. A. Rechnitz, *Anal. Chem.* **40**, 1054 (1968) and **40**, 1661 (1968).
164 W. S. Hughes, *J. Chem. Soc.* 491 (1928).
165 A. Hulanicki, R. Lewandowski and A. Lewenstam, *Anal. Chim. Acta* **110**, 197 (1979).
166 A. Hulanicki, R. Lewandowski and A. Lewenstam, *Analyst* **101**, 939 (1976).
167 A. Hulanicki and M. Trojanowicz, *Ion-Sel. El. Rev.* **1**, 207 (1979).
168 A. Hulanicki, M. Trojanowicz and M. Cichy, *Talanta* **23**, 47 (1976).
169 A. Hulanicki, M. Trojanowicz and T. Krawczynski, *Water Res.* **11**, 627 (1977).
170 A. Hulanicki, T. Krawczynski and M. Trojanowicz, *Chem. Anal. (Warsaw)* **24**, 435 (1979).
171 A. Hulanicki, A. Lewenstam and M. Maj-Zurawska, *Anal. Chim. Acta* **107**, 121 (1979).
172 M. Hukushima, H. Hukushima and T. Kuroda, *Bunseki Kagaku* **21**, 522 (1972).
173 A. A. Hussain, J. H. Kraal and H. Wahner, *J. Dent. Res.* **57**, 872 (1978).
174 N. R. Hussein, L. H. von Storp and G. G. Guilbault, *Anal. Chim. Acta* **61**, 89 (1972).
175 S. Ikeda, N. Matsuda, G. Nakagawa and K. Ito, *Denki Kagaku* **47**, 281 (1979).

176 S. Ikeda, H. Sataka, H. Hisono and T. Terezawa, *Talanta* 19, 1650 (1972).
177 B. L. Ingram, *Anal. Chem.* 42, 1825 (1970).
178 J. O. Isard, The dependence of glass-electrode properties on composition, Chapter 3 of *Glass Electrodes for Hydrogen and Other Cations* (ed. G. Eisenman), M. Dekker, New York (1969).
179 K. Irlweck and H. Soratin, *Microchim. Acta* 25 (1977).
180 K. Ito, N. Matsuda, S. Ikeda and G. Nakagawa, *Denki Kagaku* 48, 16 (1980).
181 K. Ito, N. Matsuda, T. Maeda, S. Ikeda, T. Iida and G. Nakagawa, *Denki Kagaku* 47, 220 (1979).
182 D. Jagner and V. Pavlova, *Anal. Chim. Acta* 60, 153 (1972).
183 R. Jasinski and I. Trachtenberg, *Anal. Chem.* 44, 2373 (1972).
184 R. Jasinski and I. Trachtenberg, *Anal. Chem.* 45, 1277 (1973).
185 R. Jasinski and I. Trachtenberg, *J. Electrochem. Soc.* 120, 1169 (1973).
186 R. Jasinski, I. Trachtenberg and D. Andrychuk, *Anal. Chem.* 46, 364 (1974).
187 R. Jasinski, I. Trachtenberg and G. Rice, *J. Electrochem. Soc.* 121, 363 (1974).
188 G. Johansson and K. Edstrom, *Talanta* 19, 1623 (1972).
189 J. R. Johnson, R. H. Bristow and H. H. Blau, *J. Am. Ceram. Soc.* 34, 165 (1951).
190 B. C. Jones, J. E. Heveran and B. Z. Senkowski, *J. Pharm. Sci.* 58, 607 (1969).
191 B. C. Jones, H. E. Heveran and B. Z. Senkowski, *J. Pharm. Sci.* 60, 1036 (1971).
192 M. Kataoka and T. Kambara, *Bunseki Kagaku* 23, 1157 (1974).
193 M. Kataoka and T. Kambara, *Bunseki Kiki* 10, 773 (1972).
194 M. Kataoka and T. Kambara, *Denki Kagaku* 45, 674 (1977).
195 M. Kataoka, M. Takahasi and T. Kambara, *Bunseki Kagaku* 28, 196 (1979).
196 N. A. Kazarjan and E. Pungor, *Anal. Chim. Acta* 51, 213 (1970).
197 N. A. Kazarjan and E. Pungor, *Anal. Chim. Acta* 60, 193 (1972).
198 P. J. Ke and L. W. Regier, *Anal. Chim. Acta* 53, 23 (1971).
199 P. J. Ke and L. W. Regier, *J. Fish. Res. Bd. Can.* 28, 1055 (1971).
200 D. B. Kell, P. John, M. C. Sorgato and S. J. Ferguson, *FEBS Lett.* 86, 294 (1978).
201 V. V. Kiyanskii and T. G. Aiturina, *Khim. Prom. Ser. Metody Anal. Kontrolya Kach. Prod. Khim. Prom.* 12, 38 (1980).
202 A. M. Knevel and P. F. Kehr, *Anal. Chem.* 44, 1863 (1972).
203 M. Koebel, *Anal. Chem.* 46, 1559 (1974).
204 E. Koch and C. Wagner, *Z. Physik. Chem.* (B) 38, 295 (1937).
205 T. Kojima, M. Ichise and Y. Seo, *Bunseki Kagaku* 20, 20 (1971).
206 T. Kojima, Y. Seo and J. Sato, *Bunseki Kagaku* 23, 1389 (1974).
207 T. Kojima, Y. Seo and J. Sato, *Bunseki Kagaku* 24, 772 (1975).
208 M. Komiya, *Bunseki Kagaku* 21, 911 (1972).
209 E. Körös and M. Burger, Bromide selective electrode for use in following the Zhabotinski-type oscillating chemical reaction, in *Ion-Selective Electrodes* (ed. E. Pungor), Symposium 1972, Akadémiai Kiadó, Budapest (1973), p. 191.
210 J. Koryta, *Anal. Chim. Acta* 61, 329 (1972).
211 J. Koryta, *Anal. Chim. Acta* 91, 1 (1977).
212 J. Koryta, *Anal. Chim. Acta* 111, 1 (1979).
213 J. Koryta, *Anal. Chim. Acta* 139, 1 (1982).
214 J. Koryta, J. Dvořák and V. Boháčková, *Electrochemistry*, Chapman & Hall, London (1973).
215 P. O. Kosonen, M. J. Hotokka and J. A. Mannonen, *Finnish Chem. Lett.* 90 (1978).
216 A. S. Kovalenko and L. P. Tichonova, *Teor. Eksp. Khim.* 14, 558 (1978).
217 P. A. Kryukov and S. Y. Tarasenko, *Izv. Sib. Otd. Akad. Nauk SSSR* 37, 2 (1979).
218 D. E. Lacroix and N. P. Wong, *J. Food. Prot.* 43, 672 (1980).
219 C. G. Lamm, E. H. Hansen and J. Růžička, *Anal. Lett.* 5, 451 (1972).
220 J. E. Land and C. V. Osborne, *J. Less Common Met.* 29, 147 (1972).

221 P. Lanza, *Anal. Chim. Acta* **105**, 53 (1979).
222 L. N. Lapatnicki, *Anal. Chim. Acta* **72**, 430 (1974).
223 M. J. Larsen, M. Kold and F. R. von Feher, *Caries Res.* **6**, 193 (1972).
224 S. Larsen and A. E. Widdowson, *J. Soil Sci.* **22**, 210 (1971).
225 B. C. Lawes, L. B. Fournier and O. B. Mathre, *Plating* **60**, 902 (1973).
226 B. Lengyel and E. Blum, *Trans. Faraday Soc.* **30**, 461 (1934).
227 D. A. Levaggi, W. Oyung and M. Feldstein, *J. Air Pollution Ass.* **21**, 277 (1971).
228 A. Liberti, Ion selective electrodes for heterogeneous polythene membranes, in
    Ion-Selective Electrodes (ed. E. Pungor), Symposium 1972, Akadémiai Kiadó,
    Budapest (1973), p. 37.
229 A. Liberti and M. Mascini, *Fluoride* **4**, 49 (1971).
230 T. S. Light, Industrial analysis and control with ion selective electrodes, Chapter
    10 of *Ion-Selective Electrodes* (ed. R. A. Durst), NBS Special Publication
    No. 317, Washington (1969).
231 T. S. Light, *Industrial Water Engineering* **6**, 33 (1969).
232 T. S. Light and R. F. Mannion, *Anal. Chem.* **41**, 107 (1969).
233 J. J. Lingane, *Anal. Chem.* **39**, 881 (1967).
234 J. J. Lingane, *Anal. Chem.* **40**, 935 (1968).
235 C. Liteanu, I. C. Popescu and V. Ciovirnache, *Talanta* **19**. 985 (1972).
236 O. O. Lyalin and M. S. Turaeva, *Elektrokhimiya* **3**, 256 (1977).
237 R. A. Llenado and G. A. Rechnitz, *Anal. Chem.* **44**, 468 (1972).
238 R. A. Llenado and G. A. Rechnitz, *Anal. Chem.* **44**, 1366 (1972).
239 D. A. MacInnes and M. Dole, *J. Am. Chem. Soc.* **52**, 29 (1930).
240 D. A. MacInnes and M. Dole, *Ind. Eng. Chem., Anal. Ed.* **1**, 57 (1929).
241 E. Mainka and W. Coerdt, Experiments for quick and accurate thorium assay
    by potentiometric end point determination, *Report Kernforschungszentrum
    Karlsruhe* KFK 2709 (1978).
241a V. Majer and K. Štulík, *Talanta* **29**, 145 (1982).
242 V. Majer, J. Veselý and K. Štulík, *Anal. Lett.* **6**, 577 (1973).
243 L. I. Manakova, N. N. Bausova, V. E. Moiseev, V. G. Bamburov and A. P. Sivoplyas,
    Zh. Anal. Khim. **35**, 1517 (1978).
244 A. Marbeuf, G. Demazeau, S. Turrell, P. Hagenmuller, J. Derouet and P. Caro,
    J. Solid State Chem. **3**, 637 (1971).
245 G. B. Marshall and D. Midgley, *Analyst* **103**, 438 (1978).
246 M. Mascini, *Analyst* **98**, 325 (1973).
247 M. Mascini, *Anal. Chim. Acta* **62**, 29 (1972).
248 M. Mascini and A. Liberti, *Anal. Chim. Acta* **51**, 231 (1970).
249 M. Mascini and A. Liberti, *Anal. Chim. Acta* **60**, 405 (1972).
250 M. Mascini and A. Liberti, *Anal. Chim. Acta* **64**, 63 (1973).
251 N. Matsuda, G. Nakagawu, S. Ikeda and K. Ito, *Denki Kagaku* **48**, 199 (1980).
252 J. R. McClenahen and E. Q. Schulz, *Soil Sci.* **122**, 267 (1976).
253 K. E. McLeod and H. L. Crist, *Anal. Chem.* **45**, 1272 (1973).
254 N. R. McQuaker and M. Gurney, *Anal. Chem.* **49**, 53 (1977).
255 J. Mertens, P. van den Winkel and J. Vereecken, *J. Electroanal. Chem.* **85**, 277
    (1977).
256 R. E. Mesmer, *Anal. Chem.* **40**, 443 (1968).
257 R. E. Mesmer and C. F. Baes, *Inorg, Chem.* **8**, 618 (1969).
258 V. A. Mirkin, M. A. Ilyushenko and V. V. Bakanina, *Zh. Anal. Khim.* **32**, 2282
    (1978).
259 H. H. Moeken, H. Eschrich and G. Willeborts, *Anal. Chim. Acta* **45**, 233
    (1969).
260 M. S. Mohan and G. A. Rechnitz, *Anal. Chem.* **45**, 1323 (1973).

261 G. J. Moody, N. S. Nassory, J. D. Thomas, P. Szepesváry and B. Wright, *Analyst* **104**, 1237 (1979).
262 G. J. Moody and J. D. R. Thomas, *Ion-Sel. El. Rev.* **2**, 73 (1980).
263 G. J. Moody and J. D. R. Thomas, Selective ion-sensitive electrodes, *Selected Annual Reviews of the Analytical Sciences*, (ed. L. S. Bark), Vol. 3, The Chemical Society, London (1973).
264 W. E. Morf, *The Principles of Ion-Selective Electrodes and of Membrane Transport*, Akadémiai Kiadó, Budapest, and Elsevier, Amsterdam (1981).
265 Y. Moriguchi and I. Hosokawa, *Nippon Kagaku Zasshi*, **92**, 56 (1971).
266 M. F. Mott and R. W. Guerney, *Electronic Processes in Ionic Crystals*, Dover, New York (1964), p. 31.
267 G. Muto, Y. K. Lee, K. J. Whang and K. Nozaki, *Bunseki Kagaku* **20**, 1271 (1971).
268 T. Nakano and Y. Suzuki, *Nippon Kagaku Kaishi* 1485 (1980).
268a A. Napoli, *Ann. Chim.* **68**, 443 (1978).
269 A. Napoli and M. Mascini, *Anal. Chim. Acta* **89**, 209 (1977).
270 R. Naumann and C. Weber, *Z. Anal. Chem.* **253**, 111 (1971).
271 M. Naumović, T. Bosković and O. Naumović, *Mikrochim. Acta* **2**, 537 (1977).
272 T. T. Ngo and P. O. Shargool, *Anal. Biochem.* **54**, 247 (1973).
273 E. Nicholson and E. J. Duff, *Anal. Lett.* **14**, (A 12), 887 (1981).
274 B. P. Nikolsky, *Zh. Fiz. Khim.* **10**, 495 (1937).
275 B. P. Nikolsky, *Zh. Fiz. Khim.* **27**, 724 (1953).
276 B. P. Nikolsky and M. M. Shults, *Zh. Fiz. Khim.* **36**, 1327 (1962).
277 B. P. Nikolsky, M. M. Shults, A. A. Belyustin and A. A. Lev, Recent developments in the ion-exchange theory of the glass electrode and its applications in the chemistry of glass, Chapter 6 of *Glass Electrodes for Hydrogen and Other Cations* (ed. G. Eisenman), M. Dekker, New York (1969).
278 B. P. Nikolsky and T. A. Tolmacheva, *Zh. Fiz. Khim.* **10**, 504 (1937).
279 K. Nicholson and E. J. Duff, *Analyst* **106**, 904 (1981).
280 M. Nonomura and S. Yonekura, *Kogyo Yosui* **206**, 35 (1975).
281 G. Nota, *Anal. Chem.* **47**, 763 (1975).
282 F. Oehme, *CZ-Chem.-Tech.* **3**, 27 (1974).
283 F. Oehme, *CZ-Chem.-Tech.* **4**, 183 (1975).
284 F. Oehme, *Gewässerschutz, Wasser, Abwasser* **39**, 213 (1979).
284a F. Oehme, *TAPPI Papermakers Conf. (Proc.) 1979*, 41.
285 K. Ohzeki, M. Sarukashi and T. Kambara, *Bull. Chem. Soc. Japan* **53**, 2548 (1980).
286 T. Okita, K. Kaneda, T. Yanaka and R. Sugai, *Atmos. Environ.* **8**, 927 (1974).
287 R. T. Oliver and A. G. Clayton, *Anal. Chim. Acta* **51**, 409 (1970).
288 Orion Research, *Analytical Methods Guide*, October (1972).
289 K. Oshima and N. Shibata, *Bunseki Kagaku* **23**, 392 (1974).
290 E. A. Ostrovidov, *Zavod. Lab.* **42**, 1056 (1976).
291 A. E. Owen, *J. Non-Cryst. Solids* **35–36**, 998 (1980).
292 B. Paletta and W. Bayer, *Wiener Med. Wochenschr.* **48**, 829 (1969).
293 J. Papp, *Cellul. Chem. Technol.* **5**, 147 (1971).
294 J. Papp and J. Havas, *Hung. Sci. Instr.* No. 17 (1970).
295 J. Pavel, R. Knebler and H. Wagner, *Microchem. J.* **15**, 192 (1970).
296 J. L. Penland and G. Fischer, *Metalloberfläche* **26**, 391 (1972).
297 G. A. Perley, *Anal. Chem.* **21**, 391 (1949).
298 M. H. Peters and D. M. Ladd, *Talanta* **18**, 665 (1971).
299 S. Phang and B. J. Steel, *Anal. Chem.* **44**, 2230 (1972).
300 C. S. G. Phillips and R. J. P. Williams, *Inorganic Chemistry*, Vol. 2, Oxford University Press, Oxford (1966), p. 106.
301 J. Pick, K. Tóth and E. Pungor, *Anal. Chim. Acta* **65**, 240 (1973).

302 J. Pick, K. Tóth and E. Pungor, *Anal. Chim. Acta* 61, 169 (1972).
303 A. M. Pochomis and F. D. Griffith, *Am. Ind. Hyg. Ass. J.* 32, 557 (1971).
304 H. Pokorná, *Chem. Průmsyl* 28, 238 (1978).
305 I. C. Popescu and M. Halalau, *Rev. Chim. (Bucharest)* 27, 161 (1976).
306 I. C. Popescu, E. Hopirtean, L. Savici and R. Vlad, *Rev. Roum. Chim.* 20, 993 (1975).
307 J. Portier, *Angew. Chem. Internat.* 15, 475 (1976).
308 S. Poser, W. Poser and B. Müller-Oerlinghausen, *Z. Klin. Chem. Biochem.* 12, 350 (1974).
309 R. A. Powell and M. C. Stokes, *Atmos. Environ.* 7, 169 (1973).
310 Z. Puchony, K. Tóth and E. Pungor, *Acta Chim. Acad. Sci. Hung.* 68, 177 (1971).
311 E. Pungor, *Anal. Chem.* 39, 28 A (1967).
312 E. Pungor, L. Stehli, L. André and J. György, *Anal. Chim. Acta* 46, 318 (1969).
313 E. Pungor and K. Tóth, *Analyst* 95, 625 (1970).
314 E. Pungor, K. Tóth, M. K. Pápay, L. Pólos, H. Malissa, M. Grasserbauer, E. Hake, M. F. Ebel and K. Persy, *Anal. Chim. Acta* 109, 279 (1979).
315 B. A. Raby and W. E. Sunderland, *Anal. Chem.* 39, 1304 (1967), see also correction, *Ibid.* 40, 939 (1968).
316 P. Rahlfs, *Z. Physik. Chem.* B31, 157 (1936).
317 S. Ramamoorthy, C. Guarnaschelli and D. Fecchio, *J. Inorg. Nucl. Chem.* 34, 1651 (1972).
318 G. A. Rechnitz, Analytical studies on ion-selective electrodes, Chapter 9 of *Ion-Selective Electrodes* (ed. R. A. Durst), NBS Special Publication No. 317, Washington (1969).
319 G. A. Rechnitz, *Chem. Eng. News* 45, 146 (1967).
320 G. A. Rechnitz, G. H. Fricke and M. S. Mohan, *Anal. Chem.* 44, 1098 (1972).
321 L. P. Rigdon, G. J. Moody and J. W. Frazer, *Anal. Chem.* 50, 465 (1978).
322 R. C. Rittner and T. S. Ma, *Mikrochim. Acta* 404 (1972).
323 C. J. Rix, A. M. Bond and J. D. Smith, *Anal. Chem.* 48, 1236 (1976).
324 C. E. Roberson and J. D. Hem, *Geochim. Cosmochim. Acta* 32, 1343 (1968).
325 J. W. Ross, Solid-state and liquid membrane ion selective electrodes, Chapter 2 of *Ion-Selective Electrodes* (ed. R. A. Durst) NBS Special Publication No. 317, Washington (1969).
326 J. W. Ross and M. S. Frant, *Anal. Chem.* 41, 967 (1969).
327 J. Růžička and E. H. Hansen, *Anal. Chim. Acta* 63, 115 (1973).
328 J. Růžička and C. G. Lamm, *Anal. Chim. Acta* 53, 206 (1971).
329 J. Růžička and C. G. Lamm, *Anal. Chim. Acta* 54, 1 (1971).
330 J. Růžička and O. Zajíčik, *Vod. Corp.* 19, 200 (1969).
331 J. A. Růžička and L. Mrklas, *Vod. hospod.* B23, 107 (1973).
332 R. Salomon, A. Sher and M. W. Müller, *J. Appl. Physiol.* 37, 3427 (1966).
333 J. R. Sandifer, *Anal. Chem.* 53, 312 (1981).
334 M. Sato, *Electrochim. Acta* 11, 361 (1966).
335 I. Sekerka and J. F. Lechner, *Anal. Lett.* 9, 1099 (1976).
336 I. Sekerka and J. F. Lechner, *Water Res.* 9, (1975).
337 W. Selig, *Mikrochim. Acta* 349 (1973).
338 W. Selig, *Mikrochim. Acta* 665 (1975).
339 W. Selig, *Z. Anal. Chem.* 249, 30 (1970).
339a W. Selig, *Ion-Selective Electrodes in Organic Elemental and Functional Group Analysis: A Review*, Lawrence Livermore Laboratory, University of California, Livermore, UCRL-52393 (1977).

339b W. Selig, *Ion-Selective Electrodes in Organic Elemental and Functional Group Analysis: A Review (1975 to 1978)*, Lawrence Livermore Laboratory, University of California, Livermore, UCRL-52393 Suppl. 1 (1978).
340 W. Selig and A. Salomon, *Mikrochim. Acta* 663 (1974).
341 A. R. Selmer-Olsen and A. Oien, *Analyst* 98, 412 (1973).
342 H. Schäfer, *Z. Anal. Chem.* 268, 349 (1974).
343 P. Scheide and R. A. Durst, *Anal. Lett.* 10, 55 (1977).
344 H. Schiller, *Ann. Phys.* 74, 105 (1924).
345 K. Schlyter, *Arkiv Kem* 5, 61, 73 (1953).
346 E. Schmidt and E. Pungor, *Anal. Lett.* 4, 641 (1971).
347 B. Schreiber, *Z. Anal. Chem.* 278, 343 (1976).
348 S. E. Schullery, *Chem. Phys. Lipids* 14, 49 (1975).
349 K. Schwabe, *Fortschritte der pH-Messtechnik*, VEB Verlag Technik Berlin (1958).
350 K. Schwabe and H. Dahms, *Z. Elektrochemie* 65, 518 (1961).
351 K. Schwabe and H. D. Suschke, *Angew. Chem. Internat.* 3, 36 (1964).
352 J. L. Schwartz and T. S. Light, *Totti* 53, 90 (1970).
353 I. L. Shannon and E. J. Edmonds, *Arch. Oral. Biol.* 17, (1977).
354 M. Sharp, *Anal. Chim. Acta* 59, 137 (1972).
355 M. Sharp, *Anal. Chim. Acta* 61, 99 (1972).
356 M. Sharp, *Anal. Chim. Acta* 62, 385 (1972).
357 M. Sharp, *Anal. Chim. Acta* 65, 405 (1973).
358 M. Sharp, *Anal. Chim. Acta* 76, 165 (1975).
359 M. Sharp, *Anal. Chim. Acta* 85, 17 (1976).
360 M. Sharp and G. Johansson, *Anal. Chim. Acta* 54, 13 (1971).
361 A. Shatkay, *Talanta* 17, 371 (1970).
362 D. A. Shearer and G. F. Morris, *Microchem. J.* 15, 199 (1970).
363 A. Sher, R. Solomon and K. Lee, *Physiol. Rev.* 144, 593 (1966).
364 S. Y. Shetty and R. M. Sathe, *J. Inorg. Nucl. Chem.* 39, 1838 (1977).
365 R. S. Sholtes, E. H. Meadows and S. B. Koogler, *U.S. Nat. Tech. Inform. Serv.*, PB Rep. No. 230954/OGA (1973).
365a M. M. Shults, *Dokl. AN SSSR* 194, 337 (1970).
366 L. Singer and W. D. Armstrong, *Anal. Chem.* 40, 613 (1968).
367 L. Singer and R. H. Ophang, *Clin. Med.* 25, 523 (1979).
368 I. Simonyi and I. Kálmán, Method for following the formation of aluminium chloride diisopropylate with a chloride-selective membrane electrode, in *Ion-Selective Electrodes* (ed. E. Pungor), Symposium 1972, Akadémiai Kiadó, Budapest (1973), p. 253.
369 D. R. Simpson, *Am. Mineralogist* 54, 1711 (1969).
370 J. Slanina, E. Buysman, J. Agterdenbos and G. Griepink, *Mikrochim. Acta*, 657 (1971).
371 D. Slaunwhite, J. C. Clemens and G. Reynoso, *Clin. Biochem.* 10, 44 (1977).
372 M. J. Smith and S. E. Manahan, *Anal. Chem.* 45, 836 (1973).
373 S. I. Sokoloff and A. H. Passynskii, *Z. Physik. Chem.* A160, 366 (1932).
374 R. D. Solsky and G. A. Rechnitz, *Anal. Chim. Acta* 99, 241 (1978).
375 M. H. Sorrentino and G. A. Rechnitz, *Anal. Chem.* 46, 943 (1974).
376 K. Srinivasan and G. A. Rechnitz, *Anal. Chem.* 40, 509 (1968).
377 K. Srinivasan and G. A. Rechnitz, *Anal. Chem.* 40, 1818 (1968).
378 K. Srinivasan and G. A. Rechnitz, *Anal. Chem.* 40, 1955 (1968).
379 Z. Štefanac and W. Simon, *Anal. Lett.* 1, 1 (1967).
380 Z. Štefanac and W. Simon, *Helv. Chim. Acta* 51, 74 (1968).

381 G. Stork and H. Jung, *Z. Anal. Chem.* **262**, 167 (1972).
382 E. Sucman, M. Sucmanová and O. Synek, *Z. Lebensmittel. Inters. Forsch.* **167**, 5 (1978).
383 L. Šůcha, M. Suchánek and Z. Urner, *Proc. 2nd Conf. Appl. Phys. Chem.*, Vestprém (1971), Vol. I, p. 651.
384 L. Šůcha, M. Suchánek, Z. Urner and V. Sluka, *Sborník VŠCHT Praha*, **12**, 213 (1977).
385 L. Šůcha, M. Valentová, M. Suchánek and Z. Urner, *Sborník VŠCHT Praha*, **9**, 99 (1973).
386 M. Taga, M. Mizuguchi, H. Yoshida and S. Hikime, *Bunseki Kagaku* **25**, 362 (1976).
387 S. Tanikawa, T. Adachi, N. Shiraischi, G. Nakagawa and K. Kodama, *Bunseki Kagaku* **25**, 646 (1976).
388 M. B. Terry and F. Kasler, *Mikrochim. Acta* 569 (1971).
389 H. Thompson and G. A. Rechnitz, *Chem. Instr.* **4**, 239 (1973).
390 R. J. Thompson, T. B. McMullen and G. B. Morgan, *J. Air Pollution Control Ass.* **21**, 484 (1971).
391 K. Tóth, Recent results in the field of precipitate based ion-selective electrodes, in *Ion-Selective Electrodes* (ed. E. Pungor), Symposium 1972, Akadémiai Kiadó, Budapest (1973), p. 145.
392 K. Tóth and E. Pungor, *Anal. Chim. Acta* **51**, 221 (1970).
393 A. F. Trachtenberg and I. N. Suffet, *J. Air Pollution Control Ass.* **24**, 836 (1974).
394 A. H. Truesdell and C. L. Christ, Glass electrodes for calcium, chapter 11 of *Glass Electrodes for Hydrogen and Other Cations* (ed. G. Eisenman), M. Dekker, New York (1969).
395 A. H. Truesdell and A. M. Pommer, *Science* **142**, 1292 (1962).
396 P. K. C. Tseng and W. F. Gutknecht, *Anal. Chem.* **47**, 2316 (1975).
397 P. K. C. Tseng and W. F. Gutknecht, *Anal. Chem.* **48**, 1996 (1976).
398 P. K. C. Tseng and W. F. Gutknecht, *Anal. Lett.* **9**, 795 (1975).
399 L. Tschisambou, G. Jaquen and R. Dabard, *C. R. Acad. Sci. Paris* **274**, 806 (1972).
400 J. Tušl, *Anal. Chem.* **44**, 1693 (1972).
401 J. Tušl, *Clin. Chim. Acta* **27**, 216 (1970).
402 J. Tušl, *Chem. listy* **64**, 322 (1970).
403 J. Tušl, *J. Assoc. Offic. Agr. Chemists* **53**, 267 (1970).
404 R. W. Ure, *J. Chem. Phys.* **26**, 1363 (1957).
405 V. Vajgand and V. Kalajlijeva, Determination of Pd(II) and PdI$_2$ by iodide ion-selective electrode, in *Ion-Selective Electrodes* (eds. E. Pungor and I. Buzás), Conference 1977, Akadémiai Kiadó, Budapest (1978), p. 577.
406 J. C. Van Loon, *Anal. Lett.* **1**, 393 (1968).
407 W. E. Van der Linden and R. Oostervink, *Anal. Chim. Acta* **101**, 419 (1978).
408 W. E. Van der Linden and R. Oostervink, *Anal. Chim. Acta* **108**, 169 (1979).
409 J. M. Van der Meer, G. den Boef and W. E. Van der Linden, *Anal. Chim. Acta* **76**, 261 (1975).
410 J. M. Van der Meer, G. den Boef and W. E. Van der Linden, *Anal. Chim. Acta* **85**, 317 (1976).
411 P. Venkateswarlu, *Clin. Chim. Acta* **59**, 277 (1975).
412 J. Veselý, *Coll. Czech. Chem. Comm.* **36**, 3364 (1971).
413 J. Veselý, *Coll. Czech. Chem. Comm.* **39**, 710 (1974).
414 J. Veselý, *Coll. Czech. Chem. Comm.* **46**, 368 (1981).
415 J. Veselý, *Chem. listy* **65**, 86 (1971).
416 J. Veselý, *J. Electroanal. Chem.* **41**, 134 (1973).
417 J. Veselý, O. J. Jensen and B. Nicolaisen, *Anal. Chim. Acta* **62**, 1 (1972).

418 J. Veselý and K. Štulík, *Anal. Chim. Acta* 73, 157 (1974).

419 A. E. Villa, *Analyst* 104, 545 (1979).

420 Yu. G. Vlasov, Yu. E. Ermolenko, V. V. Kolodnikov and M. S. Miloshova, *Zh. Anal. Khim.* 35, 691 (1980).

421 Yu. G. Vlasov, A. L. Grekovich, E. A. Materova, I. V. Murin and S. B. Kocheregin, *Ion. Obm. Ionometriya* 1, 170 (1976).

422 Yu. G. Vlasov and S. B. Kocheregin, *Ion. Obm. Ionometriya* 2, 243 (1979).

423 Yu. G. Vlasov, S. B. Kocheregin and Y. E. Ermolenko, *Elektrokhimiya* 13, 132 (1977).

424 Yu. G. Vlasov and S. B. Kocheregin, Structure and electrical properties of the membranes of the ion-selective electrodes based on $AgX-Ag_2S$ (X = Cl, Br, I), in *Ion-Selective Electrodes* (eds. E. Pungor and I. Buzás), Conference 1977, Akadémiai Kiadó, Budapest (1978), p. 597.

425 G. L. Vogel and W. E. Brown, *Anal. Chem.* 52, 377 (1980).

426 G. L. Vogel, L. C. Chow and W. E. Brown, *Anal. Chem.* 52, 375 (1980).

427 H. Wada and Q. Fernando, *Anal. Chem.* 43, 751 (1971).

428 C. Wagner, *J. Chem. Phys.* 21, 1819 (1953).

429 R. J. Walker and R. B. Smith, *J. Amer. Water Works Ass.* 63, 246 (1971).

430 T. B. Warner, *Anal. Chem.* 41, 527 (1969).

431 T. B. Warner, *Water Res.* 5, 459 (1971).

432 T. B. Warner and D. J. Bressan, *Anal. Chim. Acta* 63, 165 (1973).

433 T. B. Warner, M. M. Jones, G. R. Miller and D. R. Kester, *Anal. Chim. Acta* 77, 223 (1975).

434 H. F. Wastgestian, *Z. Anal. Chem.* 246, 237 (1969).

435 P. Weidenthaler and E. Pelinka, *Coll. Czech. Chem. Comm.* 34, 1482 (1969).

436 D. Weiss, *Chem. listy* 63, 1152 (1969).

437 D. Weiss, *Chem. listy* 65, 305 (1971).

438 D. Weiss, *Chem. listy* 65, 1091 (1971).

439 D. Weiss, *Chem. listy* 66, 858 (1972).

440 D. Weiss, *Chem. listy* 68, 528 (1974).

441 J. C. Westall, F. M. M. Morel and D. N. Hume, *Anal. Chem.* 51, 1792 (1979).

442 S. M. Wheeler, L. R. Fell, G. H. Fleet and R. J. Ashley, *Austr. J. Dairy Technol.* 35, 26 (1980).

443 R. J. P. Williams, *J. Chem. Soc.* 3770 (1952).

444 R. J. P. Williams, *Quart. Rev.* 24, 331 (1970).

445 F. Yamashita, T. Komatse and T. Nakagawa, *Bull. Chem. Soc. Japan* 49, 2073 (1976).

446 M. Yamazato, S. Fukuda, M. Kato and T. Yoshimori, *Denki Kagaku* 41, 789 (1976).

447 Y.-T. Yang, M. Yu, K.-Y. Mo and C.-T. Chiu, *Fen Hsi Hua Hsueh* 8, 112 (1980).

448 R. S. Yunghans and T. B. McMullen, *Fluoride* 3, 143 (1970).

449 W. H. Zachariasen, *Acta Crystallogr.* 4, 231 (1951).

450 W. H. Zachariasen, *J. Am. Chem. Soc.* 54, 3841 (1932).

451 A. Zalkin and D. H. Tompleton, *J. Am. Chem. Soc.* 75, 2453 (1953).

452 Peng-Ling Zhu, *Hua Hsueh Tung Pao* 5, 300 (1975).

453 A. F. Zhukov, A. V. Vishnyakov, Y. A. Kharif, Yu. I. Urusov, F. K. Volynets, E. I. Ryzhikov and A. V. Gordievskii, *Zh. Anal. Khim.* 30, 1761 (1975).

454 A. Zober and B. Schellmann, *Z. Klin. Chem.* 13, 197 (1975).

7

ION-SELECTIVE ELECTRODES WITH
LIQUID MEMBRANES

This group of ISEs is based on the ion-selective character of the distribution equilibrium between water and the membrane phase. As was demonstrated in chapter 3, this ion-selectivity may be affected if an ion pair is formed in the membrane (section 3.2) and increased markedly if complexes are formed in the membrane between the test ion and special complexing agents, ion carriers or ionophores (section 3.3).

7.1   Basic properties of membrane systems

The basic relationship between the selectivity coefficient and the difference between the standard Gibbs transfer energies for a system in which neither a diffusion potential in the membrane nor an ion pair in solution is formed and where the ion-exchanger ion, determinand and interferents are monovalent (see (3.2.1) and (3.2.14)) has been verified by a number of authors (frequently as a relationship between the selectivity coefficients and extraction constants or solvation energies [7, 12, 38, 151, 177, 185]). A typical dependence of this type is shown in fig. 7.1 [185], where the difference between the standard Gibbs energy for transfer of Cs^+ from water into 2-nitro-p-cymene and the Gibbs energies of cations X^+ is plotted against the logarithm of the selectivity coefficient $k^{pot}_{Cs^+, X^+}$.

The ion-pairing, affecting the selectivity coefficient according to (3.2.21), is actually the formation of a coordination compound. Consequently, specific effects appear in addition to the electrostatic interaction corresponding to the Bjerrum theory (see [106a]), according to which interactions between two oppositely charged ions increase with parameter $z_+ z_-/D$, where z_+ and z_- are the charge numbers of the cation and anion respectively, and D is the dielectric constant of the medium. This effect is important for the Ca^{2+} ISE with dialkyl-phosphate as an ion-exchanger ion in the membrane, where formation of an ion pair between the divalent cation and the ion-exchanger ion is consequently preferred over ion-pair formation with a monovalent cation. On the other hand, the effect of the solvent on the selectivity coefficient for Mg^{2+} with respect to Ca^{2+} for this kind of membrane has not yet been satisfactorily explained. For dioctylphenylphosphonate $(D \approx 6)$, $k^{pot}_{Ca^{2+},Mg^{2+}} = 10^{-2}$ while for n-decanol $(D \approx 10)$, $k^{pot}_{Ca^{2+},Mg^{2+}} \approx 1$. This levelling of the selectivity for n-decanol also appears with other ions [35].

In contrast to ISEs with neutral ion carriers in the membrane, not even qualitative rules have been formulated for the solvent effect on the behaviour of ISEs with ion-exchanger ions in a liquid membrane. A basic condition for the ion-exchanger ions is that they be strongly hydrophobic. It must hold for the standard Gibbs energy of transfer of the ion-exchanger ion X and the determinand Y that

$$\Delta G^{0,wt \to m}_{tr, X} + \Delta G^{0,wt \to m}_{tr, Y} \ll 0. \tag{7.1}$$

Consequently, an ISE for nitrate for example, a strongly hydrophilic ion, must have a strongly hydrophobic ion-exchanger ion. This conclusion has been demonstrated experimentally for a series of NO_3^- ISEs based on tetra-alkyl-ammonium salts with long alkyl chains [161] (see fig. 7.2). It was found that, in the studied series of substances, the tetradodecylammonium ion which is

Fig. 7.1. Dependence of difference between standard Gibbs transfer energies on the logarithm of the selectivity coefficient for an ISE based on 2-nitro-p-cymene. (After Scholler and Simon [185].)

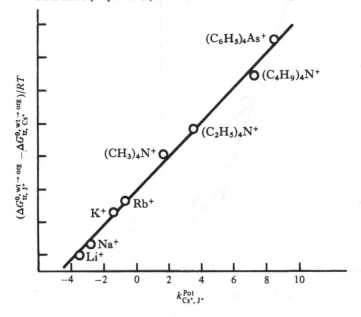

structurally similar to the ion-exchanger ion used in the NO_3^- ISE produced
by the Corning Co. [2] has the best properties. In contrast, semihydrophobic
ions, such as picrate, place less stringent demands on the ion-exchanger ion [65, 81].

It follows from study of the kinetics of transfer of ions across the phase
boundary between two immiscible electrolyte solutions (see chapter 9) that ion-
exchanger ions, where the ion is as nearly as possible symmetrically surrounded
by hydrophobic groups on all sides, are especially suitable. Amphiphilic
(amphipathic) substances, in whose molecules the hydrophobic part is separated
from the hydrophilic part, are less suitable because they have a tendency to
become adsorbed on the membrane/water phase boundary, thus retarding ion
transfer across this boundary.

For ISEs whose membranes contain ionophores, relationships derived in
section 3.3 can be successfully verified. For example, the dependence of the

Fig. 7.2. Calibration curves of NO_3^- ISE containing trioctylmethyl-(1),
tetraheptyl-(2), tetraoctyl-(3), tetradecyl-(4) and tetradodecyl-(5)
ammonium ions dissolved in dibutylphthalate. All electrodes have been
adjusted to the same $E_{0,ISE}$. (According to Nielsen and Hansen [161].)

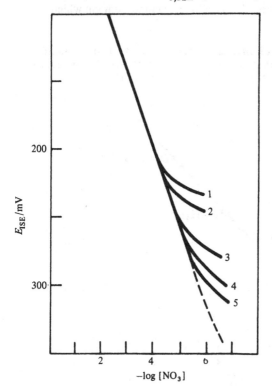

selectivity coefficient on the stability constant ratio in water (see (3.3.15) and p. 44) is depicted in fig. 7.3 for the potassium valinomycin ISE [151].

Ionophores functioning in membrane ISEs for monovalent ions were originally macrocyclic natural or synthetic substances that usually form complexes with ligand/cation composition of 1:1 (at least in cases important for ISEs). A basic property of an ionophore is that it is capable of forming a structure with a lipophilic exterior and polar cavity, as depicted in the scheme of the structure of valinomycin in fig. 7.4. The ionophore cavity must contain less than 12 and preferably 5–8 polar groups. The final complex structure must be relatively stable, which can be attained by strengthening with hydrogen bonds. It should not, however, be too rigid if ion exchange is to be sufficiently rapid [153, 193].

Newer synthetic acyclic substances that can be used in ISEs for divalent metal ions (usually forming 2:1 ligand/cation complexes) should meet the following conditions given by Simon and coworkers [2, 3]:

1. The multidentate ligand (or ligands) must, similarly to above, form a stable cavity with polar groups into which the determinand cation fits. Nonpolar groups form the outer envelope of the complex structure.

Fig. 7.3. Correlation between the selectivity coefficient and the ratio of stability constants in water for a liquid membrane ISE based on valinomycin dissolved in nitrobenzene. (After Morf [151].)

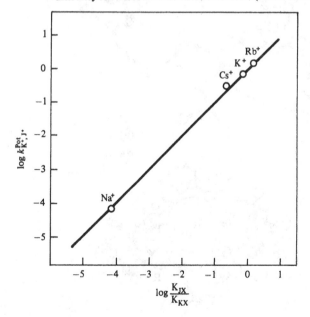

2. Coordination by oxygen atoms is preferable for the alkali metal and alkaline earth ions. Nitrogen atoms produce undesirable protonation effects.

3. The coordinating groups must be arranged in the ligand structure so that they form a five-membered ring with the central ion (Pfeiffer's rule [173]).

Fig. 7.4. Conformations of (*a*) free valinomycin and (*b*) of its potassium complex. The carbonyl oxygen atoms, P, P′, M and M′ are in especially exposed positions, so that they can initiate complexation of potassium ion. During complexation, hydrogen bonds 1 and 2 are broken, so that oxygen atoms R and R′ can take part in the co-ordination of the cation. Further smaller conformation changes allow oxygen atoms Q and Q′ to partake in formation of new hydrogen bonds, the molecule thus attaining the final round shape (see [44*a*]). (By permission of the American Association for Advancement of Science.)

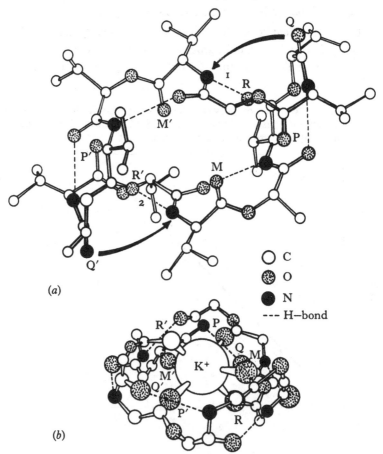

4. If the ISE is to be selective for divalent ions with respect to monovalent ions, then the ligand structure around the ion should be as 'thin' as possible.

5. An increase in the dipole moments of the polar groups increases the complex stability and the preference for divalent ions over monovalent ones.

The use of less polar solvents increases the preference for monovalent ions over divalent ones, and decreases the anion interference and the response time [154].

Complex formation between the macrocyclic ligand and the cation (originally hydrated) can be separated into two steps [153]:

1. Cation dehydration connected with the standard hydration energy of the cation multiplied by -1, $-\Delta G^0_{h,\,J^+}$; this quantity is experimentally attainable under certain conditions and it is especially easy directly to determine the difference in this quantity for a number of cations.

2. Transfer of the dehydrated cation into the cavity of the macrocyclic ligand connected with the standard Gibbs energy for the cation–ligand cavity interaction, $\Delta G^0_{i,\,J^+}$.

An approximate relationship holds between the standard Gibbs energy for complex formation ΔG^0_{JX} (and thus the stability constant K_{JX}) and quantities $\Delta G^0_{i,\,J^+}$ and $\Delta G^0_{h,\,J^+}$,

$$-RT \ln K_{JX} = \Delta G^0_{JX} = \Delta G^0_{i,\,J^+} - \Delta G^0_{h,\,J^+}. \tag{7.2}$$

The solvation interaction decreases with increasing cation radius and thus the solvation energies increase (see fig. 7.5). In contrast, the interaction energy between the cation and the ligand cavity does not change much provided the crystallographic radius of the cation is less than the cavity radius. When the cation radius is greater than the cavity radius its structure must be deformed with an increase in $\Delta G^0_{i,\,J^+}$. Because these effects act oppositely, the stability constants for the series of alkali metals usually increase at first as solvation decreases and then decrease as the decrease in the solvation is overcome by the energy consumption for ligand deformation [121, 198]. This effect is depicted in fig. 7.5 for valinomycin in acetonitrite [79].

Before the appearance of analytically useful ISEs with liquid membranes, Sollner and Shean [200] obtained a liquid ion-exchange membrane with marked selectivity for anions. The first ISE with a liquid membrane was the calcium electrode described by Ross [179] with Ca^{2+}-dialkylphenylphosphate in dioctylphenylphosphonate.

In the last ten years or so a large number of varied ion-exchanger ions have appeared, suitable for ion-selective electrodes for various cations and anions.

Hydrophobic character is attained in several ways:

1. By the presence of long alkyl chains in the molecule (tetraalkyl-ammonium or phosphonium cations, esters of phosphoric or sulphuric acid with alcohols with long alkyl chains);

2. Through the presence of aromatic or pseudoaromatic groups in the molecule (tetraphenylborate and related derivatives, the tetraphenyl-arsonium ion, cationic triphenylmethane dyes, complexes of *o*-phenanthroline with divalent nickel or ion, where the hydrophobicity is increased by further substitution by a hydrophobic group on the pseudoaromatic nucleus).

A survey of ion-exchanger ions and the corresponding determinands is given in table 7.1 for practically important systems. The reader is referred to reviews [103-105] which cover most of the literature.

Berger and coworkers [17] demonstrated the existence of macrocyclic substances capable of solubilizing alkali metal ions in nonpolar media, and described the formation of sodium and barium salts of a metabolite that had acid properties and was formed in a culture of an unspecified streptomyces. These salts were insoluble in water but dissolved in ether and benzene. The metabolite structure, originally called X 464 [17] and later nigericin [204]

Fig. 7.5. Dependence of the stability constants of valinomycin complexes of alkali metal ions on the cation radius. (After A. Hofmanová *et al.* [79].)

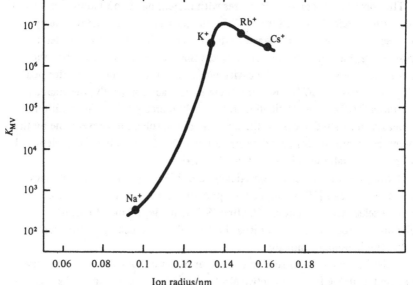

(or polyetherin [108]) is not in itself cyclic (see table 7.1, formula **XXVI**). Nonetheless, on formation of a complex with the alkali metal cation the central atom is completely surrounded by the ligand and the terminal carboxylic group (designated by an asterisk) forms a hydrogen bond with the terminal hydroxyl group (also designated by an asterisk). In this way a cyclic structure is formed.

During further research, many natural and synthetic substances with macrocyclic structures were discovered that as a rule are not charged and have the following properties in common (see [169]):

1. They form relatively stable complexes with monovalent cations, especially with alkali metal cations, where the complex stability has ion-selective character [88, 100, 193, 202, 213].

Table 7.1. *Ion-exchanger ions and ionophores used in more important ISEs.*

No.	Formula	Primary determinand	Reference
(a) *Ion-exchanging ions*			
I		Ca^{2+}	[179]
II		Ca^{2+}	[181]
III		Ca^{2+}	[147]
IV		Ca^{2+}	[19]
V		NO_3^-, BF_4^-	[180]

Table 7.1 (*cont.*)

No.	Formula	Primary determinand	Reference
VI		ClO_4^-	[180]
VII	Crystal violet	NO_3^-	[191]
VIII		ReO_4^-	[215a]
IX		Cl^-	[134]
X		NO_3^-, Cl^-	[61]
XI	Aliquat 336	ClO_4^-, Cl^-, CO_3^{2-}	[76, 83, 92]
XII		NO_3^-, HCO_3^-	[222, 223]
XIII		NO_3^-, Cl^-	[27, 52]
XIV		HCO_3^-	[60]
XV		Cs^+, quaternary ammonium ions	[16, 44]

Table 7.1 (*cont.*)

No.	Formula	Primary determinand	Reference
XVI		K⁺, acetylcholine	[13, 13a]

(b) Ionophores

XVII		Ca²⁺	[4]
XVIII	A 23187	Ca²⁺	[34]
XIX		Mg²⁺	[112]
XX		Ba²⁺	[123]

Table 7.1 (*cont.*)

No.	Formula	Primary determinand	Reference
XXI		Ba^{2+}	[2]
XXII	D-Hy-i-Valac	K^+	[51, 196]
XXIII		K^+, NH_4^+	[98, 186, 202]

$R^1 = R^2 = R^3 = R^4 = CH_3$ Nonactin

$R^1 = R^2 = R^3 = CH_3$ $R^4 = C_2H_5$ Monactin

$R^1 = R^3 = CH_3$ $R^2 = R^4 = C_2H_5$ Dinactin

$R^1 = CH_3$ $R^2 = R^3 = R^4 = C_2H_5$ Trinactin

Table 7.1 (*cont.*)

No.	Formula	Primary determinand	Reference
XXIV		K$^+$	[176]
XXV		K$^+$	[183]
XXVI	 Nigericin	Na$^+$	[204]
XXVII	 Monensin	Na$^+$	[98]
XXVIII		Na$^+$	[64]

Table 7.1 (*cont.*)

No.	Formula	Primary determinand	Reference
XXIX		Li$^+$	[62]
XXX		UO$_2^{2+}$	[190]
XXXI		H$_3$O$^+$	[118]
XXXII		Chiral tetraalkylammonium ions	[175, 209]
XXXIII		Chiral tetraalkylammonium ions	[175, 209]

2. They enable transfer of hydrophilic ions across the lipid membranes of the cells and cell organelles [6, 29a, 69, 114, 140, 148], across lipid bilayer membranes (BLM) [5, 29, 59, 72, 117, 135, 207] and across relatively thick membranes of organic solvents [155, 221].
3. They ion-selectively uncouple oxidative phosphorylation in mito-chondria [115, 140], resulting in their bactericidal action [68].
4. They lead to the formation of a membrane potential at BLM and thick membranes [30, 49, 122, 156, 193], forming a basis for their use in ion-selective electrodes, as was first demonstrated by Štefanac and Simon [202].

Neutral ion carriers in ion-selective electrodes for monovalent cations include primarily depsipeptides, which are cyclic substances composed alternatively of α-amino acid and α-hydroxy acid units. The most important member is valino-mycin (see fig. 7.4 and formula XXII in table 7.1) with a 36-membered ring, exhibiting three-fold symmetry [194]. It was prepared from cultures of *Streptomyces fulvissimus* [22, 127] and also synthetically [192]. It was first used in ISEs by Frant and Ross [51] and Simon [196]. A further group includes mostly cyclic natural or synthetic substances where the active group is an etheric oxygen. They include macrotetrolides (four-fold cyclic lactones) of the nonactin type obtained by Prelog and coworkers [42, 58] (see table 7.1, formula XXIII) from cultures of various actinomycetes, cyclic polyethers (crowns) used first in ISEs by [176], (for example XXIV and XXV) discovered by Pedersen [172] (for a review, see [206]) and finally the above-mentioned nigericin and the related monensin [1, 66] (table 7.1, formulae XXVI and XXVII). The latter substances have the character of acids and form electrically uncharged complexes. As monensin prefers sodium to potassium it is useful for Na^+ ISEs [98].

The Ba^{2+} ISE containing the acyclic membrane carrier nonylphenoxy-poly(ethylenoxy)ethanol described by Levins [123] opened new fields in ion-selective electrodes based on synthetic acyclic ionophores. Simon and coworkers synthesized several hundred substances [3] that can be used for ISEs to deter-mine alkaline earth ions, primarily Ca^{2+} [4], and also UO_2^{2+} [190], etc. (for a review, see [133, 154]). A natural ionophore for calcium is the antibiotic, A 23187 [34].

Table 7.1 surveys ionophores of practical importance for ISEs. Table 7.2 gives a survey of membrane solvents. Further sections in this chapter describe selected applications of ISEs with liquid membranes, arranged according to the determinand.

7.2 Calcium and related electrodes

A number of reviews have dealt with Ca^{2+} ISEs [144–147, 152, 197, 210]. Suitable systems for calcium electrodes are of two types, those with

Table 7.2. *Solvents used in liquid-membrane ISEs.*

No.	Formula	Name
1		Decan-1-ol
2		5-Phenylpentan-1-ol
3		1-Phenylpentan-2-ol
4		p-Butyltrifluoro-phenone
5		2-Nitro-p-cymene
6		o-Nitrophenyl-n-octyl ether
7		Dibutylsebacate
8		2-Ethylhexyl-sebacate
9		Dipentylphthalate
10		Dioctylphthalate
11		Tributylphosphate

Table 7.2 (*cont.*)

No.	Formula	Name
12		Tri-*n*-pentylphosphate
13		Tri-2-ethylhexyl-phosphate
14		Di-2-ethylhexyl-2-ethylhexyl-phosphonate
15		Dioctylphenyl-phosphonate

hydrophobic esters of phosphoric acid (I, II, III in table 7.1) as ion-exchanger ions or those with substance XVII as a membrane carrier. In the latter case the membrane must also contain Ca tetraphenylborate. Substituted diphenyl-phosphates II and III dissolved in dioctylphenylphosphonate (15 in table 7.2) or in tri-*n*-pentylphosphate, (12), were found to be better than the original dialkylphosphate I. Ion-exchanger ion IV is less important. The use of an ISE based on antibiotic A 23187 (XVIII) is hindered by marked interference from alkali metals [34].

ISEs with both these groups of systems exhibit Nernstian response to the Ca^{2+} activity in the range 10^{-1} to 10^{-7} M in metal buffers with a slope of $dE_{ISE}/d \log a_{Ca^{2+}} = 0.029$ V. They can be used in the presence of 50 mM Na^+ for down to micromolar concentrations of Ca^{2+}. The potentials of these ISEs do not depend on the hydrogen ion activity in the pH range 5.5 to 9 [146].

The most important applications of Ca^{2+} ISEs are in biomedical practice, especially in the determination of ionized calcium in serum. A pioneer work in this field was that by Moore [149] (see also [110, 111]). At present the Ca^{2+} ISE is used in the solvent polymeric version in a number of automatic devices for determining ionized calcium in serum, usually with periodic recalibration of the electrode and thermostatting to 37 °C. It should be noted that ISEs measure the activity of free Ca^{2+} in the aqueous component of the serum. This aqueous component p (ml $H_2O/100$ ml serum) is given by the relationship [218]

$$p = 99.1 - 0.73\, c(\text{protein}) - 1.03\, c(\text{lipid}), \tag{7.3}$$

where c(protein) and c(lipid) are the concentrations in g/100 ml. Of the overall calcium content, about 30-55% is bonded to the proteins as a nondialyzable component, 5-15% is bonded in complexes with low molecular weight ligands (for example HCO_3^-) and the rest is 'free'. In a healthy population, the calcium level is approximately constant [70]. In determination of calcium using an ISE it should be borne in mind that sample dilution changes not only the activity coefficients but also the equilibrium between the free and ionized calcium. The sample pH is, of course, also important as these equilibria also depend on the hydrogen ion activity; at present, the problem of whether anaerobic samples [53, 54] or samples with pH adjusted by saturating with CO_2 to a value of 7.4 [187-189] should be used is under discussion. At a constant overall calcium content, the activity of free Ca changes in certain pathological states [116]. Although the concentration of free calcium need not change, the activity changes, for example as a result of a change in the sodium concentration; thus it is necessary to define conditions for comparable activity values [97]. In addition to determination in serum, ionized calcium can also be determined in whole blood (see [21, 50, 53]). This field also includes the study of interaction between Ca^{2+} and albumin [90, 120, 184] and liberation of Ca^{2+} from mito-chondria [226].

Ion-exchangers II and III have long been used in Ca^{2+}-selective microelectrodes for intracellular and extracellular applications (for a review, see [23]), and, in spite of a certain scepticism expressed in that review, electrodes with ionophore XVII are beginning to find useful applications [63, 130, 162-165, 214].

Ca^{2+} ISEs can also be used for precise measurement of various physico-chemical data, such as the activity coefficients of Ca^{2+} [26, 125] and the stability constants of $CaATP^{2-}$ and Ca_2ATP [143].

The fact that $k_{Ca^{2+}, Mg^{2+}}^{pot} \approx 1$ for ISEs with ion-exchanger ion I and *n*-decanol (*l*) was employed by Orion Research to construct an ISE for divalent ions that can be used for determining water hardness. It has been used, for example, to measure the stability constants of ATP with Mg^{2+} and Mn^{2+} [142]. For other applications see [85].

Ligand XIX was synthesized for an ion-selective microelectrode for Mg^{2+} [112], which is useful for intracellular concentrations of magnesium greater than 0.4 mM and concentrations of 10 mM Na^+, 100 mM K^+ and 1 μM Ca^{2+}.

Suitable active substances for the membrane of the Ba^{2+} ISE are ionophore XX (nonylphenoxy-poly(ethylenoxy)ethanol, Igepal CO-880, Antarox CO-880) [123,

124] with 2-nitrophenylether as solvent [89]. This electrode is useful for titration of sulphate [94] and also yields a certain response to various polymeric uncharged alcoxylate surfactants such as Triton X-100, Tween-80, etc. [93]. An Sr^{2+} ISE based on the same principle has also been described [15].

7.3   Ion-selective electrodes for nitrates, other oxygen-containing ions and tetrafluoroborate

As a rather strongly hydrophilic anion, nitrate requires an ISE membrane containing a strongly hydrophobic cation, as described on p. 169. This function was fulfilled in the first nitrate electrode from Orion Research by cation V [180] in nitro-p-cymene 5. The electrode can be used in the pH range 4–7. In other commercial electrodes, the ion-exchanger ion is a tetra-alkylammonium salt, for example in the electrode from Corning Co., substance XIII in solvent 6 [27]. An ISE with a renewable membrane surface was found to be very useful (see section 4.1 and fig. 4.4), in which the ion-exchanger solution contains the nitrate of crystal violet VII dissolved in nitrobenzene [191]. The NO_3^- ISE also responds to nitrites that can be removed by addition of aminosulphonic acid.

Nitrate ISEs find use primarily in agriculture and environmental control. The determination of nitrates in plants [8, 137, 138, 139, 171, 199, 205] is disturbed by larger chloride concentrations which can be removed, for example by Dowex 50-X8 (in the Ag^+ form) [171]. The potentiometric method using an ISE has been found to be preferable to the nitration–distillation method based on nitration of 3,4-dimethylphenol and with the reduction–distillation method with plant samples [82]. Similarly, the ISE potentiometry is comparable to colorimetric methods for the determination of NO_3^- in soils [20, 128, 167, 182, 199]. It is recommended that samples be extracted with a $CuSO_4$ solution [167]. The determination of NO_3^- in water is described in [25, 67, 84, 201, 219]. Addition of small concentrations of fluoride permits use of the fluoride ISE as a reference electrode [129]. The Czechoslovak Ministry of Agriculture [137], on the basis of the work of the authors [82], recommends the potentiometric method with an NO_3^- ISE for the determination of nitrates in soil, plants, fodders and fertilizers after sample extraction with a solution of $AgNO_3$, $Al_2(SO_4)_3$ and $CuSO_4$. It is recommended that nitrites be removed by addition of sodium azide or amidosulphonic acid.

Nitrate ISEs can also be used for determination of nitrogen oxides after oxidation to the pentavalent form (for example with H_2O_2) [39, 41, 101, 109, 150]. The nitrate ISE system is also useful for 2,4-dichlorophenoxyacetate [73] and for tetrathionate [215].

The perchlorate ISE is based on a similar complex as the nitrate ISE, but with Fe(II) as the central atom (VI; see [180]). This system is also useful for the tetrafluoroborate ISE [180] and for the periodate ISE [48]. The tetra-fluoroborate electrode can be used for the potentiometric determination of boron after conversion to tetrafluoroborate [28], for the determination of boron carbide in Al_2O_3 [220], for analysis of borosilicate glasses [102], and for the determination of boron in silicon [113]. The periodate electrode has been used for the determination of vicinal glycols after reaction with IO_4^- [48], and for ultramicrodetermination of Mn(II) on the basis of its catalytic effect on the reaction of IO_4^- with acetylacetone [46, 47].

Similar as for nitrate electrodes, various tetra-alkylammonium ions as ion-exchanger ions (see [77, 87, 170]) are useful for perchlorate and chlorate ISEs; as ClO_4^- and ClO_3^- are less hydrophilic than NO_3^-, requirements on the hydrophobicity of the ion-exchanger ions are not as stringent.

ISEs for HCO_3^- and CO_3^{2-} [74–76, 222] are based on quaternary ammonium salts.

7.4   Ion-selective electrodes for potassium and other monovalent cations

Of the K^+ ISEs, the best properties are those of the electrode based on the ionophore valinomycin XXII [51, 196], marked by high selectivity for potassium with respect to sodium. Esters of phthalic acid (9, 10) are used as membrane solvents and it is preferable for the reasons given in section 3.3 if the membrane contains the potassium salt of hydrophobic anion XV or XVI [119, 166]. The ISE containing cyclic polyether XXV is useful for only some applications [183] because its selectivity for potassium with respect to sodium is much smaller than with the valinomycin ISE.

The most important application of the valinomycin macroelectrode is for the determination of potassium in serum [9, 126, 141, 174] and in whole blood [45, 71, 224]. This electrode with a polymeric membrane is a component of most automatic instruments for analysis of electrolytes in the serum. It has also been used for monitoring the K^+ level during heart surgery [168]. The valinomycin ISE is also useful for determination of Rb^+ [33].

The greater conductivity of an ion-exchanger solution containing the potassium salt of anion XVI in solvent 3-o-nitroxylene (Corning No. 476200) [13a] is a reason for the popularity of this system for electrophysiological measurements using ion-selective microelectrodes [216]. Application of this sensor is so widespread that only reviews and books will be mentioned, for microelectrodes in intracellular measurements [23, 32, 78, 86, 211, 217] and for the determination of K^+ in the intercellular liquid [78]. A working valinomycin microelectrode has also been constructed [166].

An ISE for acetylcholine [11] is based on the same principle (ion-exchanger solution Corning No. 476200) and has been used for determination of choline esterase [10] and the inhibitive effect of pesticides on choline esterase [14].

The lithium ISE is based on ionophore XXIX [62] in solvent 2. As the selectivity for Li^+ with respect to Na^+ is not very high, this electrode can be used only for intracellular applications as a microelectrode [212].

Sodium ISEs (especially for biological applications) contain monensin XXVII (see [98, 99, 107, 203]). Synthetic ligands for sodium ISEs were gradually developed; the latest type, XXVIII [64] has properties sufficient for the ISE to be comparable to the sodium glass electrode (see [91] for determination of Na^+ in urine).

Ammonium ISEs have been developed using a mixture of nonactin and monactin XXIII; they can be used for the determination of NH_4^+ in serum [186] (see fig. 7.6), in urine [80], in water [40, 136], as well as for kinetic study of the decomposition of arginine [159]. Unfortunately the detection limit is 50 times greater than that of the ammonium gas probe [40] (see section 4.4).

Fig. 7.6. The dependence of the potential of an ammonium ion-selective electrode to ammonium ions alone (○) and in the presence of 0.15 M Na^+, 0.005 M K^+, which is the appropriate ionic composition of blood serum (●). (After Scholer and Simon [186].)

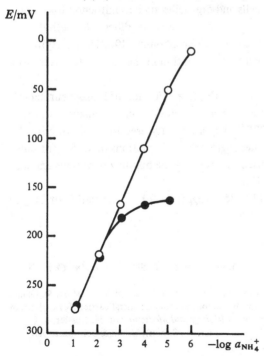

The semihydrophobic character of Cs^+ makes tetraphenylborate XV suitable as an ion-exchanger ion for the Cs ISE [16].

Bis(crown) ethers [208] have been used for the Tl(I) ISE. Where glass micro-electrodes cannot be used for pH determination because of their fragility, microelectrodes with ligand XXXI can be used for H_3O^+ [118].

Chiral ligands XXXII and XXXIII are marked by slight but clear selectivity in ISEs with respect to chiral antipodes in tetra-alkylammonium salts [175, 209].

7.5   Other systems

In recent years, experience has shown that a large number of hydrophobic ions could act as ion-exchanger ions in liquid membranes and that almost all water soluble ions not unusually hydrophilic, such as Li^+, may produce a response in such an ISE. These facts have led to an extraordinary increase in publications in this field, although relatively few of these ISEs have actually been used to solve analytical problems. This field, which is becoming rather confused, will not be discussed in this book and the reader is referred to reviews [103-105].

The solid-membrane ISE has certain disadvantages for the determination of chloride inside cells and thus ion-selective microelectrodes containing ion-exchanger Corning No. 477315 (based on a nitroxylene mixture) are used [223]. Reviews of intracellular applications of this electrode can be found in [23, 78, 86, 211, 217].

The membrane potentials of cells and organelles such as mitochondria can be measured using the distribution of hydrophobic cations (dibenzyldimethyl-ammonium [43, 44, 158, 195], tetraphenylphosphonium [96, 157], butyltri-phenylphosphonium [132]). The ion concentration in the supernatant is measured; the ion-exchanger ion is XV.

Several ISEs have been proposed for the determination of ionized surfactants; the membrane contains a surfactant ion and a suitable ion-exchanger ion [18, 31, 36, 55-57, 95, 131, 160, 178, 225]. They are mostly used for study of critical micelle concentrations and of other aggregations in solutions of surfactant salts. A surfactant ISE has also been used for study of the bonding of various anionic surfactants to bovine albumin [178].

A synthetic ionophore [190] has been suggested for determination of UO_2^{2+}.

References for Chapter 7

1 A. Agtarap, J. W. Chamberlin, M. Pinkerton and L. Steinrauf, *J. Am. Chem. Soc.* 89, 5737 (1967).
2 D. Ammann, R. Bessig, Z. Cimerman, U. Fiedler, M. Güggi, W. E. Morf, M. Oehme, H. Osswald, E. Pretsch and W. Simon, Synthetic neutral carriers for cations, in *Ion and Enzyme Electrodes in Biology and Medicine* (ed. M. Kessler, L. C. Clark, D. W. Lübbers, I. A. Silver and W. Simon), Urban & Schwarzenberg, München (1976), p. 22.

3 D. Ammann, R. Bissig, M. Güggi, E. Pretsch, W. Simon, I. J. Borowitz and L. Weiss, *Helv. Chim. Acta* **58**, 1535 (1975).

4 D. Ammann, M. Güggi, E. Pretsch and W. Simon, *Anal. Lett.* **8**, 709 (1975).

5 T. E. Andreoli, M. Tieffenberg and D. C. Tosteson, *J. Gen. Physiol.* **50**, 2527 (1967).

6 A. Azzi and G. F. Azzone, *Biochim. Biophys. Acta* **113**, 445 (1966).

7 S. Bäck and J. Sandblom, *Anal. Chem.* **45**, 1680 (1973).

8 A. S. Baker and R. Smith, *J. Agr. Food Chem.* **17**, 1284 (1969).

9 D. M. Band, J. Kratochvil, P. A. Wilson and T. Treasure, *Analyst* **103**, 246 (1978).

10 G. Baum, *Anal. Biochem.* **39**, 65 (1971).

11 G. Baum, *Anal. Lett.* **3**, 105 (1970).

12 G. Baum, *J. Phys. Chem.* **76**, 1872 (1972).

13 G. Baum, M. Lynn and F. B. Ward, *Anal. Chem. Acta* **65**, 385 (1973).

13a G. Baum and W. M. Wise, Ger. Patent U.S. No. 2024636 (1970).

14 G. Baum and F. B. Ward, *Anal. Chem.* **43**, 947 (1971).

15 E. W. Baumann, *Anal. Chem.* **47**, 959 (1975).

16 E. W. Baumann, *Anal. Chem.* **48**, 548 (1976).

17 J. Berger, A. I. Rachlin, W. E. Scott, L. H. Sternbach and M. W. Goldberg, *J. Am. Chem. Soc.* **73**, 5295 (1951).

18 B. J. Birch and D. E. Clarke, *Anal. Chim. Acta* **61**, 159 (1972).

19 R. Bloch, A. Shatkay and H. Saroff, *Biophys. J.* **7**, 865 (1967).

20 G. P. Bound, *J. Sci. Food Agric.* **28**, 501 (1977).

21 M. R. Bristow, H. D. Schwartz, G. Binetti, D. C. Harrison and J. R. Daniels, *Circul. Res.* **41**, 565 (1977).

22 H. Brockmann, M. Springorum, G. Träxler and I. Höfer, *Naturwiss.* **50**, 689 (1963).

23 H. M. Brown and J. D. Owen, *Ion-Sel. El. Rev.* **1**, 145 (1979).

24 H. M. Brown, J. P. Pemberton and J. D. Owen, *Anal. Chim. Acta* **85**, 261 (1976).

25 N. G. Bunton and N. T. Crosby, *Water Treat. Exam.* **18**, 338 (1969).

26 J. N. Butler, *Biophys. J.* **8**, 1426 (1968).

27 K. Cammann, *Working with Ion-Selective Electrodes*, Springer-Verlag, Berlin (1979).

28 R. M. Carlson and J. L. Paul, *Anal. Chem.* **40**, 1292 (1968).

29 A. Cass and A. Finkelstein, *J. Gen. Physiol.* **50**, 1765 (1967).

29a B. Chance and R. W. Estbrook, *Science* **146**, 957 (1969).

30 S. Ciani, G. Eisenman and G. Szabo, *J. Membrane Biol.* **1**, 1 (1969).

31 N. Ciocan and D. Anghel, *Tenside Deterg.* **13**, 188 (1976).

32 M. M. Civan, *Am. J. Physiol.* **234**, F 261 (1978).

33 R. F. Cosgrove and A. E. Beezer, *Anal. Chim. Acta* **105**, 77 (1979).

34 A. K. Covington and N. Kumar, *Anal. Chim. Acta* **85**, 175 (1976).

35 A. Craggs, L. Keil, C. J. Moody and J. D. R. Thomas, *Talanta* **22**, 907 (1975).

36 S. G. Cutler, P. Meares and D. G. Hall, *J. Electroanal. Chem.* **85**, 145 (1977).

37 J. E. W. Davies, G. J. Moody and J. D. R. Thomas, *Analyst* **97**, 87 (1972).

38 P. R. Danesi, F. Salvemini, G. Scibona and B. Scuppa, *J. Phys. Chem.* **75**, 554 (1971).

39 L. A. Dee, H. H. Martens, C. I. Merrill, J. T. Nakamura and F. C. Jaye, *Anal. Chem.* **45**, 1477 (1973).

40 R. Dewofs, G. Broddin, H. Clysters and H. Deelstra, *Z. Anal. Chem.* **275**, 337 (1975).

41 R. Di Martini, *Anal. Chem.* **42**, 1102 (1970).

42 J. Dominguez, J. D. Dunitz, H. Gerloch and V. Prelog, *Helv. Chim. Acta* **45**, 129 (1962).

43 P. D'Orazio and G. A. Rechnitz, *Anal. Chem.* 49, 2083 (1977).
44 L. A. Drachev, A. A. Kondrashin, V. D. Samuilov and V. P. Skulachev, *FEBS Lett.* 50, 219 (1975).
44a W. L. Duax, H. Hauptman, C. M. Weeks and D. A. Norton, *Science* 176, 911 (1972).
45 R. A. Durst, *Clin. Chim. Acta* 80, 225 (1977).
46 C. E. Efstathiou and T. P. Hadjiioannou, *Anal. Chem.* 49, 414 (1977).
47 C. E. Efstathiou and T. P. Hadjiioannou, *Talanta* 24, 270 (1977).
48 C. E. Efstathiou and T. P. Hadjiioannou, *Anal. Chem.* 47, 864 (1975).
49 G. Eisenman, S. Ciani and G. Szabo, *J. Membrane Biol.* 1, 294 (1969).
50 E. J. Fogt, A. R. Eddy, A. H. Clemens, J. Fox and H. Heath, *Clin. Chem.* 26, 1425 (1980).
51 M. S. Frant and J. W. Ross, *Science* 167, 987 (1970).
52 C. Fuchs, *Ionenselektive Elektroden in der Medizin*, Thieme, Stuttgart (1976).
53 C. Fuchs, D. Dorn, C. McIntosh, F. Scheler and B. Kroft, *Clin. Chim. Acta* 67, 99 (1976).
54 Ch. Fuchs and Ch. McIntosh, *Clin. Chem.* 23, 610 (1977).
55 T. Fujinaga, S. Okazaki and H. Freizer, *Anal. Chem.* 46, 1842 (1974).
56 K. Fukamachi and N. Ishibashi, *Bunseki Kagaku* 27, 152 (1978).
57 C. Gavach and C. Bertrand, *Anal. Chim. Acta* 55, 385 (1971).
58 H. Gerloch and V. Prelog, *Liebig's Ann. Chem.* 669, 121 (1963).
59 V. A. Gotlib, E. P. Buzhinskii and A. A. Lev, *Biofizika* 13, 562 (1968).
60 A. L. Grekovich, E. A. Materova and N. V. Garbuzova, *Zh. Anal. Khim.* 28, 1206 (1973).
61 A. L. Grekovich, E. A. Materova and G. I. Shchekina, *Elektrokhimiya* 10, 342 (1974).
62 M. Güggi, U. Fiedler, E. Pretsch and W. Simon, *Anal. Lett.* 8, 857 (1975).
63 M. Güggi, M. Kessler, F. Greitschus, V. Wiegand and W. Meesmann, Measurement of extracellular ion activities (potassium ion, sodium ion, calcium ion) during acute coronary acclusion, in *Frontiers of Biological Energy* (ed. P. L. Dutton, J. S. Leigh and A. Scarpa), Vol. 2, Academic Press, New York (1978). p. 1427.
64 M. Güggi, M. Oehme, E. Pretsch and W. Simon, *Helv. Chim. Acta* 58, 2417 (1876).
65 T. P. Hadjiioannou and E. P. Diamandis, *Anal. Chim. Acta* 94, 443 (1977).
66 M. E. Haney and M. M. Hoehn, *Antimicrobiol. Agents Chemotherapy* 1967, 349.
67 E. H. Hansen, A. K. Ghose and J. Růžička, *Analyst* 102, 705 (1977).
68 F. M. Harold and J. R. Baarda, *J. Bacteriol.* 94, 53 (1967).
69 E. J. Harris and B. C. Pressman, *Nature* 216, 918 (1967).
70 E. K. Harris and D. L. DeMets, *Clin. Chem.* 17, 983 (1971).
71 J. Havas, L. Kecskés and R. Somodi, *Hung. Sci. Instr.* 41, 47 (1977).
72 D. A. Haydon and S. B. Hladky, *Quart. Rev. Biophysics* 5, 187 (1972).
73 N. Hazemoto, N. Kamo and Y. Kobatake, *J. Assoc. Off. Anal. Chem.* 59, 1097 (1976).
74 H. B. Herman and G. A. Rechnitz, *Anal. Chim. Acta* 76, 155 (1975).
75 H. B. Herman and G. A. Rechnitz, *Anal. Lett.* 8, 147 (1975).
76 H. B. Herman and G. A. Rechnitz, *Science* 184, 1074 (1974).
77 K. Hiiro, G. J. Moody and J. D. R. Thomas, *Talanta* 22, 918 (1975).
78 P. Hník, E. Syková, N. Kříž and F. Vyskočil, Determination of ion activity changes in excitable tissues with ion-selective microelectrodes, Chapter 5 of *Medical and Biological Applications of Electrochemical Devices* (ed. J. Koryta), John Wiley and Sons, Chichester (1980).
79 A. Hofmanová, J. Koryta, M. Březina, T. H. Ryan and K. Angelis, *Inorg. Chim. Acta* 37, 135 (1979).
80 J. H. C. Hoge, H. J. A. Hazenberg and C. H. Gips, *Clin. Chim. Acta* 55, 273 (1974).

81 D. Homolka, *Coll. Czech. Chem. Comm.* **44**, 3644 (1979).
82 J. Hubáček and K. Bernatzik, *Chem. Listy* **70**, 513 (1976).
83 A. Hulanicki and R. I. Lewandowski, *Chem. Anal. (Warsaw)* **19**, 53 (1974).
84 A. Hulanicki and M. Trojanowicz, *Ion-Sel. El. Rev.* **1**, 207 (1979).
85 A. Hulanicki and M. Trojanowicz, *Chem. Anal. (Warsaw)* **18**, 235 (1973).
86 *Ion-Selective Microelectrodes and Their Use in Excitable Tissues* (ed. E. Syková, P. Hník and L. Vyklický), Plenum Press, New York (1981).
87 N. Ishibashi, A. Jyo and K. Matsumoto, *Chem. Lett.* 1297 (1973).
88 V. T. Ivanov, I. A. Laine, N. D. Abdulayev, L. B. Senyavina, E. M. Popov, Yu. A. Ovchinnikov and M. M. Shemyakin, *Biochem. Biophys. Res. Commun.* **34**, 803 (1969).
89 A. M. Y. Jaber, G. J. Moody and J. D. R. Thomas, *Analyst* **101**, 179 (1976).
90 J. S. Jakobs, R. S. Hattner and D. S. Bernstein, *Clin. Chim. Acta* **31**, 467 (1971).
91 H. B. Jenny, D. Ammann, R. Döring, B. Magyar, R. Asper and W. Simon, *Microchim. Acta* II, 125 (1980).
92 H. J. James, G. P. Carmack and H. Freiser, *Anal. Chem.* **44**, 856 (1972).
93 D. L. Jones, G. J. Moody and J. D. R. Thomas, *Analyst* **106**, 439 (1981).
94 D. L. Jones, G. J. Moody, J. D. R. Thomas and M. Hangos, *Analyst* **104**, 973 (1979).
95 K. M. Kale, E. L. Cussler and D. F. Evans, *J. Phys. Chem.* **84**, 593 (1980).
96 N. Kamo, M. Muratsugu, R. Hongah and Y. Kobatake, *J. Membrane Biol.* **49**, 105 (1979).
97 R. A. Kaufman and N. W. Tietz, *Clin. Chem.* **26**, 640 (1980).
98 O. Kedem, E. Loebel and M. Furmansky, German Patent No. 2027128 (23.12.1970).
99 M. Kessler, J. Höfer and B. A. Krumme, *Anesthesiology* **45**, 184 (1976).
100 B. T. Kilbourn, J. D. Dunitz, L. A. R. Pioda and W. Simon, *J. Mol. Biol.* **30**, 559 (1967).
101 B. M. Kneebone and H. Freiser, *Anal. Chem.* **45**, 449 (1973).
102 R. L. Kochen, *Anal. Chim. Acta* **71**, 451 (1974).
103 J. Koryta, *Anal. Chim. Acta* **91**, 1 (1977).
104 J. Koryta, *Anal. Chim. Acta* **111**, 1 (1979).
105 J. Koryta, *Anal. Chim. Acta* **139**, 1 (1982).
106 J. Koryta, J. Dvořák and V. Boháčková, *Electrochemistry*, Chapman & Hall, London (1973).
107 K. Kotera, N. Satake, M. Honda and M. Fujimoto, *Membrane Biochem.* **2**, 323 (1979).
108 T. Kubota, S. Matsutani, M. Shiro and H. Koyama, *Chem. Comm.* 1541 (1968).
109 D. Kuroda, *Bunseki Kagaku* **22**, 1191 (1973).
110 J. H. Ladenson and G. N. Browers, *Clin. Chem.* **19**, 565 (1973).
111 J. H. Ladenson and G. N. Browers, *Clin. Chem.* **19**, 575 (1973).
112 F. Lanter, D. Erne, D. Ammann and W. Simon, *Anal. Chem.* **52**, 2400 (1980).
113 P. Lanza and P. L. Buldini, *Anal. Chim. Acta* **75**, 149 (1975).
114 H. A. Lardy, S. N. Graven and S. Estrada-O., *Fed. Proc.* **26**, 1355 (1967).
115 H. A. Lardy, D. Johnson, W. C. McMurray, *Arch. Biochem. Biophys.* **78**, 587 (1958).
116 L. Larson and S. Ohman, *Clin. Chem.* **24**, 1962 (1978).
117 P. Läuger, *Science* **178**, 24 (1972).
118 O. H. Le Blanc, J. F. Brown, J. F. Klebe, L. W. Niedrach, G. M. J. Slusarczuk and W. H. Stoddard, *J. Appl. Physiol.* **40**, 644 (1976).
119 O. H. Le Blanc, Jr and W. T. Grubb, *Anal. Chem.* **48**, 1658 (1976).
120 C. E. Leme and H. B. Silva, *Clin. Chim. Acta* **77**, 287 (1977).

121 J.-M. Lehn, *Struct. Bonding* 16, 1 (1973).
122 A. A. Lev and E. P. Buzhinskii, *Cytologia* 9, 106 (1967).
123 R. J. Levins, *Anal. Chem.* 43, 1047 (1971).
124 R. J. Levins, *Anal. Chem.* 44, 1544 (1972).
125 J. V. Leyendekkers and M. Whitfield, *J. Phys. Chem.* 75, 957 (1971).
126 J. A. Lustgarten, R. E. Wenk, C. Byrd and B. Hall, *Clin. Chem.* 20, 1217 (1974).
127 J. C. MacDonald and G. P. Slater, *Canad. J. Biochem.* 46, 573 (1968).
128 M. K. Mahendrappa, *Soil Sci.* 132 (1969).
129 S. E. Manahan, *Anal. Chem.* 42, 128 (1970).
130 E. Marban, T. J. Rink, R. W. Tsien and R. Y. Tsien, *Nature* 286, 845 (1980).
131 C. McCallum and P. Meares, *Electrochim. Acta* 19, 537 (1974).
132 J. E. G. McCarthy and S. J. Ferguson, *Biochem. J.* 196, 311 (1981).
133 P. C. Meier, D. Ammann, W. E. Morf and W. Simon, Liquid-membrane ion-selective electrodes and their biomedical applications, Chapter 2 of *Medical and Biological Application of Electrochemical Devices* (ed. J. Koryta), John Wiley and Sons, Chichester (1980).
134 P. C. Meier, D. Ammann, H. F. Osswald and W. Simon, *Med. Progr. Technol.* 5, 1 (1977).
135 *Membranes – A Series of Advances* (ed. G. Eisenman), Vol. 2, M. Dekker, New York (1973).
136 J. Mertens, P. Van den Winkel and D. L. Massart, *Anal. Lett.* 6, 81 (1973).
137 *Methods for Introduction of Innovations into Practice 23: Determination of nitrates in soil, plants and fodder,* (in Czech), Institute of Scientific and Technological Information for Agriculture, Prague (1979).
138 P. J. Milham, *Analyst* 95, 758 (1970).
139 P. J. Milham, A. S. Awad, R. E. Paull and J. H. Bull, *Analyst* 95, 751 (1970).
140 P. Mitchell, *Chemiosmotic Coupling and Energy Transduction*, Glynn Research, Bodmin (1968).
141 D. S. Miyada, K. Inami and G. Matsuyama, *Clin. Chem.* 17, 27 (1971).
142 M. S. Mohan and G. A. Rechnitz, *Arch. Biochem. Biophys.* 162, 194 (1974).
143 M. S. Mohan and G. A. Rechnitz, *J. Am. Chem. Soc.* 94, 1714 (1972).
144 G. J. Moody and J. D. R. Thomas, *Anal. Chem. Symp. Ser.* 2, *Electroanal., Hyg., Environ., Clin., Pharm. Chem.*, University of Wales Inst. Sci. Technol., Cardiff (1980), p. 11.
145 G. J. Moody and J. D. R. Thomas, Design, principles and behaviour of sensitive calcium ion-selective electrodes in *Electroanalysis in Hygiene, Environmental, Clinical and Pharmaceutical Chemistry* (ed. W. F. Smith), Elsevier, Amsterdam (1980), p. 11.
146 G. J. Moody and J. D. R. Thomas, *Proc. Anal. Div. Chem. Soc.* 16, 32 (1979).
147 G. J. Moody and J. D. R. Thomas, *Ion-Sel. El. Rev.* 1, 3 (1979).
148 C. Moore and B. C. Pressman, *Biochem. Biophys. Res. Comm.* 15, 562 (1964).
149 E. W. Moore, Studies with ion-exchange calcium electrodes in biological fluids: Some applications in biomedical research and clinical medicine, Chapter 7 of *Ion-Selective Electrodes* (ed. R. A. Durst), NBS Special Publication No. 317, Washington (1969).
150 G. P. Morie, C. J. Ledford and C. A. Glower, *Anal. Chim. Acta* 60, 397 (1972).
151 W. E. Morf, *The Principles of Ion-Selective Electrodes and of Membrane Transport*, Akadémiai Kiadó, Budapest (1981).
152 W. E. Morf, M. Oehme and W. Simon, Response characteristics of K^+, Ca^{2+} and other liquid membrane electrodes, in *Ionic Actions in Vascular Smooth Muscle* (ed. E. Betz), Springer-Verlag Berlin (1976), p. 1.
153 E. Morf and W. Simon, *Helv. Chim. Acta* 54, 2683 (1971).

154 W. E. Morf and W. Simon, Ion-selective electrodes based on neutral carriers, Chapter 3 of *Ion-Selective Electrodes in Analytical Chemistry* (ed. H. Freiser), Vol. 1, Plenum Press, New York (1978).

155 W. E. Morf, P. Wuhrmann and W. Simon, *Anal. Chem.* 48, 1031 (1976).

156 P. Mueller and D. O. Rudin, *Biophys. Biochem. Res. Comm.* 26, 398 (1967).

157 M. Muratsugu, N. Kamo, Y. Kobatake and K. Kimura, *Bioelectrochem. Bioenerget.* 6, 493 (1979).

158 M. Muratsugu, N. Kamo, K. Kurihara and Y. Kobatake, *Biochim. Biophys. Acta* 464, 613 (1977).

159 T. A. Neubecker and G. A. Rechnitz, *Anal. Lett.* 5, 653 (1972).

160 J. E. Newbery and V. Smith, *Colloid Polymer Sci.* 256, 494 (1978).

161 H. J. Nielsen and E. H. Hansen, *Anal. Chim. Acta* 85, 1 (1976).

162 C. Nicholson, *Fed. Proc.* 39, 1519 (1980).

163 C. Nicholson, *Trends Neurosci. (Pers. Ed.)* 3, 216 (1980).

164 C. Nicholson, G. Ten Bruggencate, H. Stöckle and R. Steinberg, *J. Neurophysiol.* 41, 1026 (1978).

165 C. Nicholson, G. Ten Bruggencate, R. Steinberg and H. Stöckle, *Proc. Nat. Acad. Sci.* 74, 1287 (1977).

166 M. Oehme and W. Simon, *Anal. Chim. Acta* 86, 21 (1976).

167 A. Øien and A. R. Selmer-Olsen, *Analyst* 94, 888 (1969).

168 H. F. Osswald, R. Asper, W. Dimai and W. Simon, *Clin. Chem.* 25/1, 39 (1979).

169 Ju. A. Ovchinnikov, V. T. Ivanov and M. M. Shkrob, *Membrane-active Complexones*, Elsevier, Amsterdam, (1974).

170 E. Paglia-Bubin, T. Mussini and R. Galli, *Z. Naturforsch. A*, 26, 154 (1971).

171 J. L. Paul and R. M. Carlson, *J. Agr. Food Chem.* 16/5, 766 (1968).

172 C. J. Pedersen, *J. Am. Chem. Soc.* 89, 7017 (1967).

173 P. Pfeiffer, *Angew. Chem.* 53, 93 (1940).

174 L. A. R. Pioda, W. Simon, H.-R. Bosshard and H. Ch. Curtius, *Clin. Chim. Acta* 29, 289 (1970).

175 V. Prelog, *Pure Appl. Chem.* 50, 893 (1978).

176 G. A. Rechnitz and E. Eyal, *Anal. Chem.* 44, 370 (1972).

177 R. E. Reinsfelder and F. A. Schultz, *Anal. Chim. Acta* 65, 425 (1973).

178 H. M. Rendall, *J. Chem. Soc. Faraday Trans. I*, 72, 481 (1976).

179 J. W. Ross, *Science* 156, 1378 (1967).

180 J. W. Ross, Solid-state and liquid membrane ion-selective electrodes, Chapter 2 of *Ion-Selective Electrodes* (ed. R. A. Durst), NBS Special Publication, No. 317, Washington (1969).

181 J. Růžička, E. H. Hansen and J. C. Tjell, *Anal. Chim. Acta* 67, 155 (1973).

182 J. Růžička, E. H. Hansen and E. A. Zagatto, *Anal. Chim. Acta* 88, 1 (1977).

183 O. Ryba and J. Petránek, *J. Electroanal. Chem.* 44, 425 (1973).

184 C. Sachs and A. M. Bourdeau, *Clin. Orthop.* 78, 24 (1971).

185 R. Scholer and W. Simon, *Helv. Chim. Acta* 55, 801 (1972).

186 R. P. Scholer and W. Simon, *Chimia* 24, 372 (1970).

187 H. D. Schwartz, *Clin. Chem.* 22, 461 (1976).

188 H. D. Schwartz, *Clin. Chem.* 23, 610 (1977).

189 H. D. Schwartz, *Clin. Chim. Acta* 64, 227 (1975).

190 J. Senkýř, D. Ammann, P. C. Meier, W. E. Morf, E. Pretsch and W. Simon, *Anal. Chem.* 51, 786 (1979).

191 J. Senkýř and J. Petr, *Chem. listy* 73, 1097 (1979).

192 M. M. Shemyakin, N. A. Aldanova, E. I. Vinogradova and M. J. Feigina, *Tetrahedron Let.* 1921 (1963).

193 M. M. Shemyakin, J. A. Ovchinnikov, V. T. Ivanov, V. K. Antonov, E. I. Vino-
    gradova, A. M. Shkrob, G. G. Malenkov, A. V. Yevstratov, I. A. Laine,
    J. I. Melnik and I. D. Ryabova, *J. Membrane Biol.* 1, 402 (1969).
194 M. M. Shemyakin, E. I. Vinogradova, M. Yu. Feignina, N. A. Aldanova,
    N. F. Loginova, I. D. Ryabova and I. A. Pavlenko, *Experientia* 21, 548 (1965).
195 T. Shinbo, N. Kamo, K. Kurikara and Y. Kobatake, *Arch. Biochem. Biophys.*
    187, 414 (1978).
196 W. Simon, Electrochemical Society Meeting, New York, May 1969.
197 W. Simon, D. Ammann, M. Oehme and W. E. Morf, *Ann. New York Acad. Sci.*
    307, 52 (1978).
198 W. Simon, W. E. Morf and P. Ch. Meier, *Struct. Bonding* 16, 113 (1973).
199 G. R. Smith, *Anal. Lett.* 8, 503 (1975).
200 K. Sollner and G. M. Shean, *J. Am. Chem. Soc.* 86, 1901 (1964).
201 T. G. Sommerfeldt, R. A. Milne and G. C. Kozub, *Commun. Soil Sci. Plant Anal.*
    2, 415 (1971).
202 Z. Štefanac and W. Simon, *Chimia* 20, 436 (1966).
203 O. K. Stefanova, V. F. Gorshkova and E. A. Materova, *Ion. Obm. Ionometriya*
    2, 183 (1979).
204 L. K. Steinrauf, M. Pinkerton and J. W. Chamberlin, *Biochem. Biophys. Res.
    Comm.* 33, 29 (1968).
205 A. W. M. Sweetsur and A. G. Wilson, *Analyst* 100, 485 (1975).
206 *Synthetic Multidentate Macrocyclic Compounds* (ed. R. M. Izatt and J. J.
    Christensen), Academic Press, New York (1978).
207 G. Szabo, G. Eisenman and S. Ciani, Ion distribution equilibria in bulk phases and
    the ion transport properties in lipid bilayer membranes produced by neutral
    macrocyclic antibiotics, *Proc. Coral Gables Conf. on the Physical Principles of
    Biological Membranes*, 1968, Gordon and Breach, New York (1969).
208 H. Tamura, K. Kimura and T. Shono, *J. Electroanal. Chem.* 115, 115 (1980).
209 A. P. Thoma, Z. Cimerman, V. Fiedler, D. Bedenković, M. Güggi, P. Jordan,
    K. May, E. Pretsch, V. Prelog and W. Simon, *Chimia* 29, 344 (1975).
210 J. D. R. Thomas, *Lab. Pract.* 27, 857 (1978).
211 R. C. Thomas, *Ion-Sensitive Intracellular Microelectrodes. How to Make and
    Use Them*, Academic Press, London, (1978).
212 R. C. Thomas, W. Simon and M. Oehme, *Nature* 285, 754 (1975).
213 D. C. Tosteson, P. C. Cook, T. E. Andreoli and M. Tieffenberg, *J. Gen. Physiol.*
    50, 2513 (1967).
214 R. Y. Tsien and T. J. Rink, *Biochim. Biophys. Acta* 599, 623 (1980).
215 O. H. Tuovinen and J. D. Nicholas, *App. Environment. Microbiol.* (1977), p. 477.
215a Yu. I. Urosov, V. V. Sergievskii, A. Ya. Syrchenkov, A. F. Zhukov and A. V.
    Gardieoskii, *Zh. Anal. Khim.* 30, 1757 (1975).
216 J. L. Walker, *Anal. Chem.* 43, 89A (1971).
217 J. L. Walker, Single cell measurement with ion-selective electrodes, Chapter 4 of
    Medical and Biological Application of Electrochemical Devices (ed. J. Koryta),
    John Wiley and Sons, Chichester (1980).
218 W. H. Waugh, *Metabolism* 18, 706 (1969).
219 D. Weiss, *Chem. listy* 69, 202 (1975).
220 H. E. Wilde, *Anal. Chem.* 45, 1526 (1973).
221 H. K. Wipf, A. Olivier and W. Simon, *Helv. Chim. Acta* 53, 1605 (1970).
222 W. M. Wise, US Patent 3 723 281 (1972).
223 W. M. Wise, US Patent No 3 801 486 (1972).
224 H. R. Wuhrmann, *Biomed. Tech., (Stuttgart)*, 21, 191 (1976).
225 A. Yamauchi, T. Kunisaki, T. Minematsu, Y. Tomokiyo, T. Yamaguchi and
    H. Kimizuka, *Bull. Chem. Soc. Japan* 51, 2791 (1978).
226 R. K. Yamazaki, D. L. Mickey and M. Stary, *Anal. Biochem.* 93, 430 (1979).

8

POTENTIOMETRIC BIOSENSORS

This group of sensors involves composite systems (see section 3.4), consisting of an internal ion-selective electrode (usually a glass electrode) and a hydrophilic layer, in which a certain component of the test solution (usually, but not always, the determinand) is converted into a form sensed by the internal ISE. The reacting systems are mostly components of an enzymic reaction, with the enzyme suitably immobilized in the hydrophilic layer. The biochemical reaction can also take place directly in cells or in cell models, liposomes, which are immobilized in the layer surrounding the ISE, or, finally, in a thin layer cut from a tissue and attached to the ISE (section 8.2). The layer is either directly in contact with the ISE, for example with the enzyme electrodes (section 8.1), or is placed so that the test solution in a flow-through system first contacts the layer of the immobilized enzyme and the products formed are then led to the ISE; this is an electrode with an enzyme reactor [29] (see fig. 8.1). Enzymic reactions can, of course, also take place in homogeneous phases with an ISE indicating the concentration of the products, but this case will not be dealt with here.

Fig. 8.1. An enzyme electrode (*a*) and an enzyme reactor electrode (*b*). (After Morf [38].)

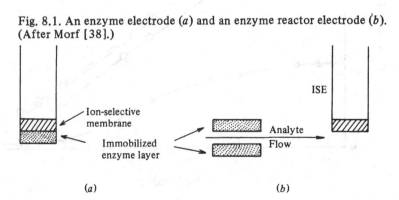

(*a*) (*b*)

8.1   Enzyme electrodes

The potentialities offered by chemical processes catalyzed by enzymes immobilized in a polymeric matrix are obvious and are now successfully utilized in various ways [6, 13, 16, 17, 40, 43, 44, 50, 58, 61]. This idea was introduced into electroanalytical chemistry by Clark and Lyons [9], who proposed a glucose electrode, with glucose oxidase immobilized between cuprophane membranes and with amperometric determination of the hydrogen peroxide formed by the reaction

$$\text{glucose} + O_2 \xrightarrow{\text{glucose oxidase}} \text{gluconic acid} + H_2O_2. \tag{8.1}$$

This system is also the most important enzyme electrode in practice.

Among potentiometric enzyme sensors, the urea enzyme electrode is the oldest (and the most important). The original version consisted of an enzyme layer immobilized in a polyacrylamide hydrophilic gel and fixed in a nylon netting attached to a Beckman 39137 glass electrode, sensitive to the alkali metal and NH_4^+ ions [19, 24]. Because of the poor selectivity of this glass electrode, later versions contained a nonactin electrode [20, 22] (cf. p. 187) and especially an ammonia gas probe [25] (cf. p. 72). This type of urea electrode is suitable for the determination of urea in blood and serum, at concentrations from 5 to 0.05 mm. Figure 8.2 shows the dependence of the electrode response

Fig. 8.2. Dependence of the urea-ISE potential on the urea concentration at four urease contents in the membrane gel: ● 8.0, ○ 140, △ 175, □ 233 mg cm^{-3}. (After Guilbault and Nagy [20]. By permission of the American Chemical Society.

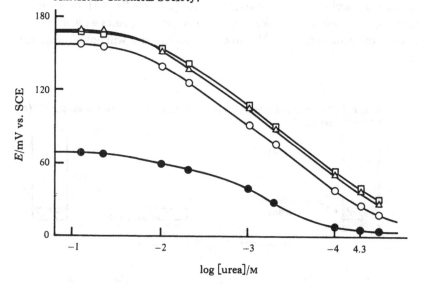

on the urea concentration at various concentrations of the enzyme in the immobilized layer.

The amygdaline electrode [47] that does not find extensive practical application is suitable as a model for investigation of the properties of enzyme electrodes in general [36], because of its specific response. The electrode is based on the reaction

$$C_6H_5 . \underset{\substack{|\\ O . C_{12}H_{21}O_{10}}}{CH} . CN \xrightarrow{\beta\text{-glucosidase}} 2 C_6H_{12}O_6 + C_6H_5 CHO + HCN, \qquad (8.2)$$

$$O . C_{12}H_{21}O_{10} + 2H_2O$$

which attains a maximum rate at pH = 12.7. An iodide ISE monitors the concentration of the cyanide ions formed. Other enzyme electrodes have been developed for L-amino acids [18], phenylalanine [21, 27], glutamine [23], creatinin [54], 5'-adenosine monophosphate [42], nitrite [35], nitrate [34], acetyl choline [10, 12, 32, 56], methionine [14], o-acetyl-L-serine [41] and histidine [59].

Enzyme electrodes with amperometric indication have certain advantages over potentiometric sensors, chiefly because the product of the enzymic reaction is consumed at the electrode and thus the response time is decreased. For this reason, the potentiometric glucose enzyme electrode, based on reaction (8.1) followed by the reaction of H_2O_2, with iodide ions sensed by an iodide ISE [39], has not found practical use.

The process occurring in the immobilized enzyme layer is the enzymic reaction of substrate (determinand) S with enzyme E,

$$S + E \underset{k_-}{\overset{k_+}{\rightleftharpoons}} SE \xrightarrow{k_2} E + nP, \qquad (8.3)$$

together with diffusion of the substrate in this layer toward the internal ISE and diffusion of the product both toward the ISE and out from the immobilized layer. The rate of the enzymic reaction proper is expressed by the Michaelis and Menthen equation,

$$\frac{d[P]}{dt} = -n \frac{d[S]}{dt} = nk_2[E] \frac{[S]}{K_M + [S]}, \qquad (8.4)$$

where $K_M = k_+/k_-$ is the Michaelis constant. Because the concentration of the active enzyme, [E], may depend on the concentrations of activators or inhibitors of the enzymic reaction, enzyme electrodes can, in principle, be used for the determination of these substances [17]. The dependence of the enzyme electrode potential on the determinand concentration and the response time depend on all the above processes. Therefore the thickness of the enzyme layer and the enzyme concentration in the membrane are important parameters. The resultant response times are rather long, usually several minutes, and thus enzyme electrodes are not particularly suitable for determinations such as flow-injection analysis. The problem of the response of potentiometric electrodes was also solved theoretically; however, the solutions [7, 8, 26, 37, 38, 55] are poorly suited to interpretation of experimental results, except for Morf's work.

8.2   Other biosensors

Bacterial electrodes [11, 31, 33, 46, 48, 49, 60] In this type of electrode, a suspension of suitable bacteria is placed between the sensor proper and a dialysis membrane that prevents passage of high-molecular substances (see fig. 8.3). The sensor is usually a gas probe. In the simple types of bacterial electrode, the determinand is converted by a suitable strain of bacteria into a product sensed by the gas probe. Thus it is possible to determine arginine [46], glutamine [48], L-aspartic acid [31], L-histidine [60] and nitrate [33]. Hybrid bacterial – enzyme electrodes contain both a bacterial strain and a suitable enzyme. For example, an extract from fungus *Neurospora chossa* can be used as a source of NAD^+ nucleosidase and an *Escherichia coli* culture as a source of nicotinamide deaminase, so that the electrode responds to NAD^+ [49] as a result of the series of reactions

$$NAD^+ + H_2O \xrightarrow{\text{NAD nucleosidase}}$$

$$\text{nicotinamide} + \text{adenosine diphosphateribose}$$

$$\text{nicotinamide} + H_2O \xrightarrow[\text{deaminase}]{\text{nicotinamide}} \text{nicotinic acid} + NH_3 .$$

A potentiometric determination of lysozyme is based on a system similar to a bacterial electrode [11]. The cells of the bacteria, *Micrococcus lysodeicticus*, readily accept trimethylphenyl ammonium ions ($TMPA^+$) from the solution. Lysozyme decomposes the cell membranes and $TMPA^+$ is liberated. The rate

Fig. 8.3. A bacterium electrode. (After D'Orazio *et al.* [11].)

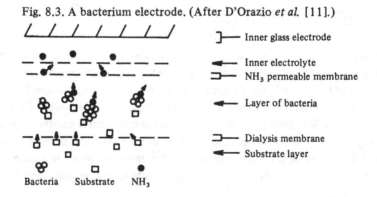

]—	Inner glass electrode
◄—	Inner electrolyte
⊐—	NH_3 permeable membrane
◄—	Layer of bacteria
⊐—	Dialysis membrane
◄—	Substrate layer

Bacteria   Substrate NH_3

of the liberation, determined by the TMPA$^+$ ISE, corresponds to the lysozyme activity. The lifetime of these sensors is rather short; an optimistic estimate is up to two weeks.

Tissue electrodes [2, 3, 4, 5, 45, 57]. In these biosensors, a thin layer of tissue is attached to the internal sensor. The enzymic reactions taking place in the tissue liberate products sensed by the internal sensor. In the glutamine electrode [5, 45], a thick layer (about 0.05 mm) of porcine liver is used and in the adenosine-5'-monophosphate electrode [4], a layer of rabbit muscle tissue. In both cases, the ammonia gas probe is the indicator electrode. Various types of enzyme, bacterial and tissue electrodes were compared [2]. In an adenosine electrode a mixture of cells obtained from the outer (mucosal) side of a mouse small intestine was used [3]. The stability of all these electrodes increases in the presence of sodium azide in the solution that prevents bacterial decomposition of the tissue. In an electrode specific for the antidiuretic hormone [57], toad bladder is placed over the membrane of a sodium-sensitive glass electrode. In the presence of the antidiuretic hormone, sodium ions are transported through the bladder and the sodium electrode response depends on the hormone concentration.

Immunosensors Monitoring of the antibody reaction with antigen, which is the basis of immunoanalysis, involves the determination of excess Ag$^+$ with Ag$_2$S ISE after the addition of AgNO$_3$ to sulphur-containing proteins [1]. The method is based on monitoring the precipitine reaction of the antibody with the antigen (see p. 139).

The function of an immuno-electrode [28] containing a model antibody (Concanavalin A) fixed in a polymeric film on a platinum electrode is probably based on other effects than those utilized in ISEs. Immuno-electrodes suitable for direct determination of antibodies were prepared by fixing a conjugate of an ionophore and an immunogen (for example the compound of dibenzo-18-crown-6 with dinitrophenol) in a PVC membrane. This system responded to the antibody against dinitrophenol [52, 53].

This group also includes immuno-analysis using liposomes, containing a liquid antigen in the membrane and tetrapentylammonium ion (TPA$^+$) inside. The antibody causes immunolysis of the liposome and liberation of TPA$^+$, which is determined by an ISE [51].

The determination of the lysozyme activity using the bacterial electrode (see p. 139) can also be used for immuno-analysis of biotine and avidine. The determination is based on the inhibition reaction of avidine with the biotine-lysozyme conjugate. After the reaction, the conjugate is no longer capable of dissolving the cell wall of the bacteria. The determination of biotine is similar [15].

References for Chapter 8

1 P. W. Alexander and G. A. Rechnitz, *Anal. Chem.* **46**, 1253 (1974).
2 M. A. Arnold and G. A. Rechnitz, *Anal. Chem.* **52**, 1170 (1980).
3 M. A. Arnold and G. A. Rechnitz, *Anal. Chem.* **53**, 515 (1981).
4 M. A. Arnold and G. A. Rechnitz, *Anal. Chem.* **53**, 1837 (1981).
5 M. A. Arnold and G. A. Rechnitz, *Anal. Chim. Acta* **113**, 351 (1980).
6 T. Ya. Bart and E. A. Materova, *Ion. Ob. Ionometriya* **2**, 73 (1979).
7 J. E. Brody and P. W. Carr, *Anal. Chem.* **52**, 977 (1980).
8 P. W. Carr, *Anal. Chem.* **49**, 799 (1977).
9 L. C. Clark and C. Lyons, *Ann. New York Acad. Sci.* **102**, 29 (1962).
10 K. L. Crochet and J. C. Montalvo, *Anal. Chim. Acta* **66**, 269 (1973).
11 P. D'Orazio, M. E. Meyerhoff and G. A. Rechnitz, *Anal. Chem.* **50**, 1531 (1978).
12 P. Durand, A. David and D. Thomas, *Biochim. Biophys. Acta* **527**, 277 (1978).
13 M. M. Fishman, *Anal. Chem.* **52**, 185 R (1980).
14 K. W. Fung, S. S. Kuan, H. Y. Sung and G. G. Guilbault, *Anal. Chem.* **51**, 2319 (1979).
15 C. R. Gebauer and G. A. Rechnitz, *Anal. Biochem.* **103**, 180 (1980).
16 G. G. Guilbault, *Enzyme Microb. Technology* **2**, 258 (1980).
17 G. G. Guilbault, Use of enzyme electrodes in biomedical investigations, Chapter 9 of *Medical and Biological Applications of Electrochemical Devices* (ed. J. Koryta), John Wiley and Sons, Chichester (1980).
18 G. G. Guilbault and E. Hrabánková, *Anal. Chem.* **42**, 1779 (1970).
19 G. G. Guilbault and J. Montalvo, *J. Am. Chem. Soc.* **92**, 2533 (1970).
20 G. G. Guilbault and G. Nagy, *Anal. Chem.* **45**, 417 (1973).
21 G. G. Guilbault and G. Nagy, *Anal. Lett.* **6**, 301 (1973).
22 G. G. Guilbault, G. Nagy and S. S. Kuan, *Anal. Chim. Acta* **67**, 195 (1973).
23 G. G. Guilbault and F. R. Shu, *Anal. Chim. Acta* **56**, 333 (1971).
24 G. G. Guilbault, R. K. Smith and J. G. Montalvo, *Anal. Chem.* **41**, 600 (1969).
25 G. G. Guilbault and W. Stokbro, *Anal. Chim. Acta* **76**, 237 (1975).
26 H. F. Hameka and G. Rechnitz, *Anal. Chem.* **53**, 1586 (1981).
27 C. P. Hsiung, S. S. Kuan and G. G. Guilbault, *Anal. Chim. Acta* **90**, 45 (1977).
28 J. Janata, *J. Am. Chem. Soc.* **97**, 2914 (1975).
29 G. Johansson and L. Ögren, *Anal. Chim. Acta* **84**, 23 (1976).
30 R. K. Kobos, Potentiometric enzyme methods, in *Ion-Selective Electrodes in Analytical Chemistry* (ed. H. Freiser), Vol. 2, Plenum Press, New York (1980), p. 1.
31 R. Kobos and G. A. Rechnitz, *Anal. Lett.* **10**, 751 (1977).
32 R. Kobos and G. Rechnitz, *Arch. Biochem. Biophys.* **175**, 11 (1976).
33 R. K. Kobos, D. J. Rice and D. S. Fournoy, *Anal. Chem.* **51**, 1122 (1979).
34 C. H. Kiang, S. S. Kuan and G. G. Guilbault, *Anal. Chem.* **50**, 1319 (1978).
35 C. H. Kiang, S. S. Kuan and G. G. Guilbault, *Anal. Chim. Acta* **80**, 209 (1975).
36 M. Mascini and A. Liberti, *Anal. Chim. Acta* **68**, 177 (1974).
37 L. D. Mell and J. T. Maloy, *Anal. Chem.* **47**, 299 (1975).
38 W. E. Morf, *Mikrochim. Acta* **II**, 317 (1980).
39 G. Nagy, L. H. von Storp and G. G. Guilbault, *Anal. Chim. Acta* **66**, 443 (1973).
40 T. T. Ngo, *Int. J. Biochem.* **11**, 459 (1980).
41 T. Ngo and P. Shargool, *Anal. Biochem.* **54**, 247 (1973).
42 D. S. Papastathopoulos and G. A. Rechnitz, *Anal. Chem.* **48**, 862 (1976).
43 G. A. Rechnitz, *Bio-Selective Membrane Electrodes in Trace Organic Analysis*, NBS Special Publication No. 519, U.S. Government Printing Office, Washington, D.C., (1979), p. 525.

44 G. A. Rechnitz, *Science* **214**, 287 (1981).
45 G. A. Rechnitz, M. A. Arnold and M. E. Meyerhoff, *Nature* **278**, 466 (1979).
46 G. A. Rechnitz, R. K. Kobos, S. J. Riechel and C. R. Gebauer, *Anal. Chim. Acta* **94**, 357 (1977).
47 G. A. Rechnitz and R. Llenado, *Anal. Chem.* **43**, 283 (1971).
48 G. A. Rechnitz, T. L. Riechel, R. K. Kobos and M. E. Meyerhoff, *Science* **199**, 440 (1978).
49 T. L. Riechel and G. A. Rechnitz, *J. Membrane Sci.* **4**, 243 (1978).
50 F. Scheller and D. Pfeiffer, *Z. Chemie* **18**, 50 (1978).
51 K. Shiba, Y. Umezawa, T. Watanabe, S. Ogawa and S. Fujiwara, *Anal. Chem.* **52**, 1610 (1980).
52 R. L. Solsky and G. A. Rechnitz, *Anal. Chim. Acta* **123**, 135 (1981).
53 R. L. Solsky and G. A. Rechnitz, *Science* **204**, 1308 (1979).
54 H. Thompson and G. A. Rechnitz, *Anal. Chem.* **46**, 246 (1974).
55 C. Tran-Minh and J. Beaux, *Anal. Chem.* **51**, 91 (1979).
56 C. Tran-Minh, R. Guyonet and J. Beaux, *C. R. Acad. Sci. Paris*, **C286**, 115 (1978).
57 S. Updike and I. Treichel, *Anal. Chem.* **51**, 1643 (1979).
58 P. Vadgema, Enzyme Electrodes, in *Ion-Selective Electrode Methodology* (ed. A. K. Covington) Vol. 2, CRC Press, Boca Raton, Fla., (1979), p. 23.
59 R. R. Wallers, P. A. Johnson and R. P. Buck, *Anal. Chem.* **52**, 1684 (1980).
60 R. R. Walters, B. E. Moriarty and R. P. Buck, *Anal. Chem.* **52**, 1680 (1980).
61 T.-M. Yuan, *Fen Hsi Hua Hsuech* **7**, 149 (1979).

9

VOLTAMMETRY AT THE INTERFACE OF TWO IMMISCIBLE ELECTROLYTE SOLUTIONS

The previous chapters dealt with ISE systems at zero current, i.e. at equilibrium or steady-state. The properties of the interface between two immiscible electrolyte solutions (ITIES), described in sections 2.4 and 2.5, will now be used to describe a dynamic method based on the passage of electrical current across ITIES. Voltammetry at ITIES (for a survey see [3, 8, 9, 10, 11, 12, 18]) is an inverse analogue of potentiometry with liquid-membrane ISEs and thus forms a suitable conclusion to this book.

The properties of ITIES are in many respects analogous to those of the metal electrode/electrolyte solution interface [8]. This analogy is especially pronounced when electrolyte $J_1 X_1$, dissolved in one phase (an organic solvent virtually immiscible with water), is strongly hydrophobic, whereas electrolyte $J_2 X_2$, dissolved in the other phase (water), is strongly hydrophilic. This situation is characterized by the following inequalities for the distribution coefficients of the substances:

$$K_{J_1 X_1}^{wt,0} \gg 1, \quad K_{J_2 X_2}^{wt,0} \ll 1. \tag{9.1}$$

Fig. 9.1. A four-electrode system for polarization of ITIES.

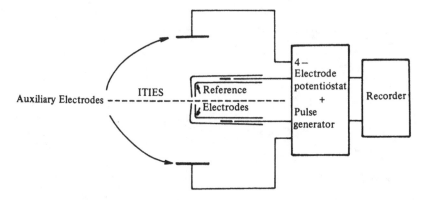

Under these circumstances [12, 13], the ITIES behaves as an ideally polarizable interface, i.e. within a certain range of electrical potential values between water and the organic phase, $\Delta_o^{wt}\phi$ attains the value applied from an external source. In other words, the ITIES behaves analogously to a mercury electrode in a KCl solution. Here it is also possible to change this potential by varying the voltage applied from an external source, within a certain potential range ('potential window'). Similar to the mercury electrode, where the boundaries of the potential window are determined by the properties of the calomel electrode and the potassium amalgam electrode, the boundaries for the ITIES are the potentials at which ions J_1^+ and X_1^- are transferred from the organic into the aqueous phase, and ions J_2^+ and X_2^- from the aqueous into the organic phase. The limits of the potential window are characterized by the values of the standard potentials for the transfer of the appropriate ions from water into the organic solvent (2.2.10). However, similar to metal electrodes, transfer of ions J_1^+, J_2^+, X_1^- and X_2^- is possible at lower potentials than the standard potential of the transfer and thus the width of the potential window decreases. This narrowing depends on the concentration of the given electrolyte.

Fig. 9.2. The electrolytic cell for polarization of a stationary ITIES: RE_1, RE_2 – reference electrodes; CE_1, CE_2 – auxilliary electrodes; L_1, L_2 – Luggin capillaries; F_1, F_2 – sintered glass, wt – aqueous phase, org – organic phase.

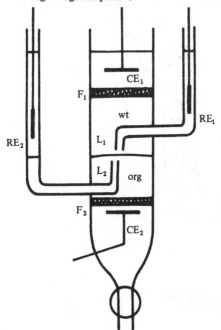

If one of the solvents contains ion K^+ in addition to ions J_1^+, J_2^+, X_1^- and X_2^-, and this ion is present at a substantially lower concentration than the latter ions and has a standard potential for transfer from water into the organic solvent lying inside the potential window, then, at a suitable applied potential difference, the ion is transferred from one phase into the other. This is a phenomenon quite analogous to those encountered in polarography and related methods. As the standard rate constants for transfer of ions across ITIES (cf. (2.5.1)) are large, most cases, especially those important in analyses, correspond to reversible processes at current values corresponding to ion transfer. The activity of ions close to the ITIES is then given by (2.2.11) and the magnitude of the current is determined by the diffusion from the interior of one phase toward the ITIES and from the ITIES toward the interior of the other phase. Electrolytes $J_1 X_1$ and $J_2 X_2$ then act as base electrolytes.

Experimental arrangement [1, 5, 6, 13, 14, 15, 17, 19, 20] In contrast to electro-chemical methods utilizing metal electrodes, the ohmic drop of the potential

Fig. 9.3. The electrolyte dropping electrode: RE_1, RE_2 – reference electrodes; CE_2 – auxilliary electrode; 1 – aqueous phase formed by dropping of the electrode; 2 – nitrobenzene phase; 3 – Teflon capillary; 4 – sintered glass. The arrow denotes the connection to the reservoir with the aqueous electrolyte and the auxiliary electrode CE_1. (After Samec *et al.* [20].)

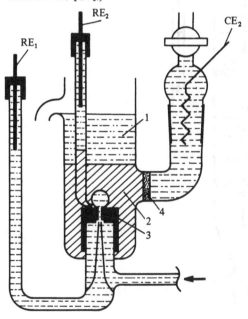

in the two solutions must be compensated for in electrolysis at ITIES. Therefore, a four-electrode system must be used (fig. 9.1). Two Luggin capillaries connecting reference electrodes RE_1 and RE_2 are placed as close as possible to the ITIES, while two auxiliary electrodes CE_1 and CE_2 are separated by frits from the test solutions [1, 17]. As reference electrodes, silver chloride electrodes in aqueous solutions are mostly used. The reference electrode for the aqueous phase usually contains an alkali metal chloride, whereas the reference electrode for the organic phase contains the chloride of a cation identical with the cation of the base electrolyte in the organic phase. An interface is then formed between the Luggin capillary and the organic phase, with a defined potential difference given by (2.2.11) for this cation (fig. 9.2).

A dropping electrolyte electrode (fig. 9.3) was also developed for the study of electrolysis at ITIES. When the density of the organic phase is greater than that of the aqueous phase (for example, as with nitrobenzene), the aqueous drops move upwards.

A hanging electrolyte drop electrode (fig. 9.4) was constructed for analytical purposes requiring electrolytic preconcentration of the test component [14, 15].

Methods of polarizing ITIES [1, 13, 14, 15, 17, 19, 20] In the beginning of the study of ITIES, chronopotentiometry was used as the most easily accessible

Fig. 9.4. The hanging electrolyte drop electrode. Symbols are the same as in fig. 9.2. (After Mareček and Samec [15].)

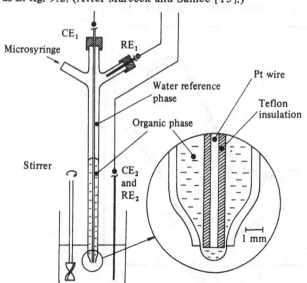

Fig. 9.5. Polarogram with aqueous dropping electrode: 1 – base electrolyte, water: 0.05 M LiCl, 1 M $MgSO_4$; nitrobenzene: 0.05 M tetrabutylammonium tetraphenylborate; 2 – 1 mM tetramethylammonium chloride added to the aqueous phase. The dashed curve was obtained by subtraction of 1 from 2. (After Samec *et al.* [20].)

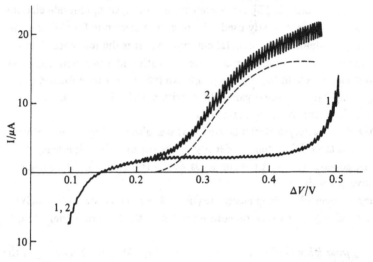

Fig. 9.6. Cyclic voltammogram of 1 mM Cs^+ in water at polarization rate 5 mV/s (full line). Aqueous phase: 50 mM LiCl; nitrobenzene phase: 50 mM tetrabutylammonium tetraphenylborate. Dotted curve: base electrolyte, dashed curve: voltammogram of Cs^+ corrected for the base electrolyte current. (After Samec *et al.* [19].)

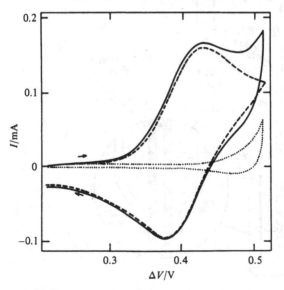

method [1]. Although theoretical works dealing with electrolysis at ITIES and based on chronopotentiometry still appear (see for example [16]), voltammetry with triangular voltage pulses has undoubted advantages for analytical applications [17]. However, this method requires a four-electrode potentiostat (fig. 9.6). Among commercial instruments, the Wenking LB 75 potentiostat is suitable for the purpose. In addition to potential-pulse voltammetry, polarography can also be used, yielding results analogous to those from the classical Heyrovský method [13, 20] (fig. 9.5). However, work with a dropping electrolyte electrode is rather difficult.

A high sensitivity is attained by using pre-electrolysis into a hanging electrolyte drop, followed by a series of pulses used in differential pulse polarography [15].

Solvents, base electrolytes As solvents for the organic phase, liquids with relative dielectric constants above 10 are suitable. So far, nitrobenzene has been used almost exclusively, but dichloroethane can also be employed [7]. Tetrabutylammonium tetraphenylborate [1] is the most commonly used base

Fig. 9.7. Determination of acetylcholine by differential pulse polarography with hanging electrolyte drop electrode. Acetylcholine concentrations: 0–0, 1–0.5 ppm, 2–1 ppm, 3–2 ppm, 4–5 ppm. (After Mareček and Samec [15].)

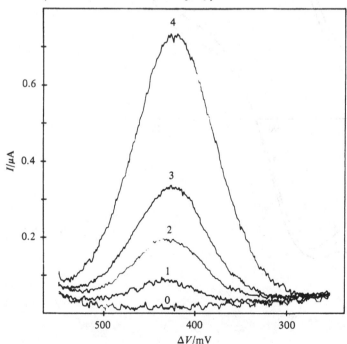

electrolyte for the non-aqueous phase and LiCl for the aqueous phase [13].
To widen the potential window, even more hydrophobic ions should be used
in the organic phase, for example tetraphenylarsonium [15] or crystal violet [21]
cations and 1,2-π-dicarbalkylcobaltate(III) anions [13].

Voltammetry at ITIES in the analysis The study of simple ion transfer has
involved the transfer of tetramethylammonium and tetraethylammonium
ion [1, 13, 20], cesium ion [19], acetylcholine [15, 22], picrate ion [4],

Fig. 9.8. Cyclic voltammogram of 0.75 mM nonactin dissolved in
nitrobenzene. Base electrolytes: 82 mM NaCl (aqueous phase), 50 mM
tetrabutylammonium tetraphenylborate (nitrobenzene). Polarization rate
50 mV s^{-1}. 1 – original curve; 2 – after subtraction of the base electrolyte
current (After Homolka *et al.* [3].)

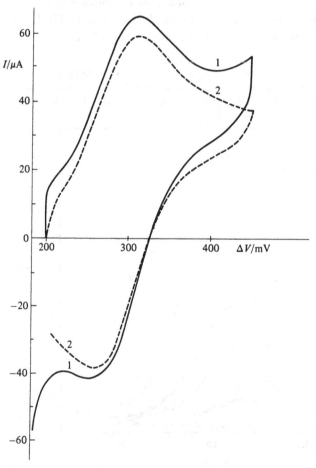

n-octanate, laurylsulphate, rhodanide and perchlorate [21]. The half-wave
potential, obtained by various electrochemical methods, is

$$\Delta V_{1/2} = \Delta_0^{wt}\phi_i^0 + (RT/2z_iF)\ln(D_{wt}/D_0),\qquad(9.2)$$

where D_{wt} and D_0 are the diffusion coefficients in the aqueous and organic
phases, respectively. Using simple voltammetry with a triangular pulse (fig. 9.6),
ions can be determined down to a concentration of 5×10^{-5} mol dm^{-3}. With
pre-electrolysis and differential pulse polarography [15], down to 0.5 ppm
acetylcholine could be determined in an aqueous solution (fig. 9.7).

Another analytically useful phenomenon in electrolysis at ITIES is ion
transfer facilitated by ionophores present in the non-aqueous phase [8]. If
the ionophore is present at a low concentration in the non-aqueous phase and
the aqueous phase contains a large concentration of the cation that is bound
in a complex with the ionophore (for example as a component of the base
electrolyte), then a voltammetric wave controlled by diffusion of the ionophore
toward the ITIES or by diffusion of the complex formed away from the ITIES
into the bulk of the organic phase appears at a potential lower than the potential
of simple cation transfer. The peak height of this wave is proportional to the
ionophore concentration in the solution and can be used for the determination
(fig. 9.8). This effect has been observed with valinomycin, nonactin, cyclic
polyethers and other substances [2, 3, 23]. The half-wave potential of these waves is

$$\Delta V_{1/2} = \Delta_0^{wt}\phi_K^0 - \frac{RT}{z_KF}\ln[K_{KY}a_K(wt)] + \frac{RT}{z_KF}\ln\frac{D_Y^0}{D_{KY}^0}$$

$$\approx \Delta_0^{wt}\phi_K^0 - \frac{RT}{z_KF}\ln[K_{KY}a_K(wt)],\qquad(9.3)$$

where $\Delta_0^{wt}\phi_K^0$ is the standard potential difference for transfer of ion K with
charge number z_K between water and the organic phase, K_{KY} is the stability
constant for the complex of ion K with ionophore Y and $a_K(wt)$ is the activity
of ion K in the aqueous phase [8].

The coccidiostat monensin (Table 7.1, XXVII) dissolved in nitrobenzene with
a dilute HNO$_3$ + NaCl solution as aqueous phase electrolyte can be accurately
determined by this method, also in *Streptomyces* culture extracts [11a].

References for Chapter 9

1 C. Gavach and F. Henry, *J. Electroanal. Chem.* 54, 361 (1974).
2 A. Hofmanová, Le Q. Hung and M. W. Khalil, *J. Electroanal. Chem.* 135, 257 (1982).
3 D. Homolka, Le Q. Hung, A. Hofmanová, M. W. Khalil, J. Koryta, V. Mareček, Z. Samec, S. K. Sen, P. Vanýsek, J. Weber, M. Březina, M. Janda and I. Stibor, *Anal. Chem.* 52, 1606 (1980).

4 D. Homolka and V. Mareček, *J. Electroanal. Chem.* **112**, 91 (1980).
5 Le Q. Hung, *Chem. listy* **74**, 1089 (1980).
6 Le Q. Hung and P. Vanýsek, *Chem. listy* **74**, 869 (1980).
7 Z. Koczorowski and G. Geblewicz, *J. Electroanal. Chem.* **108**, 117 (1980).
8 J. Koryta, *Electrochim. Acta* **24**, 293 (1979).
9 J. Koryta, Electrolysis at the interface of two immiscible electrolyte solutions, in Ion-Selective Electrodes, *Proceedings of the 3rd Symposium, 1980* (ed. E. Pungor and E. Buzás), Akadémiai Kiadó, Budapest (1981), p. 53.
10 J. Koryta, *Hung. Sci. Instr.* No. 49, 25 (1980).
11 J. Koryta, M. Březina, A. Hofmanová, D. Homolka, Le Q. Hung, M. W. Khalil, V. Mareček, S. K. Sen, P. Vanýsek and J. Weber, *Bioelectrochem. Bioenerget.* **7**, 61 (1980).
11a J. Koryta, W. Ruth, P. Vanýsek and A. Hofmanová, *Anal. Lett.* **15B**, 1685 (1982).
12 J. Koryta and P. Vanýsek, Electrochemical phenomena at the interface of two immiscible electrolyte solutions, in Advances in Electrochemistry and Electrochemical Engineering (ed. H. Gerischer and C. W. Tobias), Wiley-Interscience, New York (1981), p. 113.
13 J. Koryta, P. Vanýsek and M. Březina, *J. Electroanal. Chem.*, **75**, 211 (1977).
14 V. Mareček and Z. Samec, *Anal. Chim. Acta*, **141**, 65 (1982).
15 V. Mareček and Z. Samec, *Anal. Lett.* **14 (B15)**, 1241 (1981).
16 O. R. Melroy and R. P. Buck, *J. Electroanal. Chem.*, **136**, 19 (1982).
17 Z. Samec, V. Mareček, J. Koryta and M. W. Khalil, *J. Electroanal. Chem.* **83**, 393 (1977).
18 Z. Samec, V. Mareček, P. Vanýsek and J. Koryta, *Chem. listy* **74**, 715 (1980).
19 Z. Samec, V. Mareček and J. Weber, *J. Electroanal. Chem.* **117**, 841 (1979).
20 Z. Samec, V. Mareček, and J. Weber and D. Homolka, *J. Electroanal. Chem.* **99**, 385 (1979).
21 P. Vanýsek, *J. Electroanal. Chem.* **121**, 149 (1984).
22 P. Vanýsek and M. Behrendt, *J. Electroanal. Chem.* **130**, 287 (1981).
23 P. Vanýsek, W. Ruth and J. Koryta, *J. Electroanal. Chem.* in press.

APPENDIX
A LIST OF SELECTIVITY COEFFICIENTS

Data taken mainly from E. Pungor, K. Tóth and A. Hrabeczy-Páll, *Pure and Applied Chemistry*, **51**, 1915 (1979), and P. C. Meier, D. Ammann, W. E. Morf and W. Simon, Liquid-membrane ion-selective electrodes and their biomedical applications, Chapter 2 of *Medical and Biological Applications of Electrochemical Devices* (ed. J. Koryta), John Wiley & Sons, Chichester (1980). MS - mixed solution technique, SS - separate solution technique; symbols for ion-exchanging ions from table 7.1.

Determined	Electrode characteristics	Interferents	$\log k_{J,K}^{pot}$	Method of determination
Cl^-	AgCl precipitate	CrO_4^{2-}	-4.3	Calculated from P_{JA}
		PO_4^{3-}	-3.9	
		CO_3^{2-}	-4.2	
		AsO_4^{3-}	-3.5	
		CrO_4^{2-}	-4.3	MS
		CO_3^{2-}	-4.3	
		PO_4^{3-}	-4.3	
Cl^-	XI in decanol	NO_3^-	0.3	
		SO_4^{2-}	-1.4	
		Br^-	-0.5	
Br^-	$AgBr/Ag_2S$ (Orion)	Cl^-	-3.0	MS
		I^-	3.0	
		SCN^-	-3.0	
		S^{2-}	10.0	
I^-	AgI precipitate	Br^-	-4.0	MS
		Cl^-	-5.4	
		F^-	-5.3	
		SCN^-	0.0	
		SO_4^{2-}	-5.4	
		NO_3^-	-5.0	

Appendix (continued)

Determined	Electrode characteristics	Interferents	$\log k_{J,K}^{pot}$	Method of determination
S^{2-}	Ag_2S	CN^-	-3.2	MS
		I^-	-7.8	
		all others	<-10.7	MS
F^-	LaF_3 single crystal	OH^-	-1.0	MS
		all others	<-3.0	
Cd^{2+}	$CdS+Ag_2S$	Mg^{2+}	-3.8	SS
		Ca^{2+}	-3.7	
		Zn^{2+}	-3.4	
		Co^{2+}	-1.7	
		Ni^{2+}	-1.5	
		Al^{3+}	-0.9	
		H^+	0.4	
		Mn^{2+}	0.4	
		Pb^{2+}	0.8	
		Tl^+	2.1	
		Fe^{2+}	2.3	
Pb^{2+}	$PbS+Ag_2S$	Ni^{2+}	-3.5	MS
		Mn^{2+}	-3.9	
		Zn^{2+}	-4.5	
		Ni^{2+}	-2.4	
		Cd^{2+}	-0.5	SS
ClO_4^-	VI (Orion)	I^-	-1.5	MS
		NO_3^-	-2.4	
		OAc^-	-2.8	
		Br^-	-3.0	
		HCO_3^-	-3.1	
		F^-	-3.5	
NO_3^-	V (Orion)	ClO_4^-	3.0	
		I^-	1.3	
		ClO_3^-	0.3	
		Br^-	-0.9	
		SH^-	-1.4	
		NO_2^-	-1.4	
		CN^-	-2.0	
		HCO_3^-	-2.0	
		Cl^-	-2.4	
		OAc^-	-3.4	
		CO_3^{2-}	-3.7	
NO_3^-	VII in nitrobenzene	ClO_4^-	3.0	MS
		SCN	1.8	
		I^-	1.3	
		Br^-	-0.9	
		NO_2	-1.3	
		Cl^-	-2.3	
		HCO_3^-	-2.4	
		$H_2PO_4^-$	-4.0	
		SO_4^{2-}	-4.7	

Appendix (continued)

Determined	Electrode characteristics	Interferents	$\log k_{J,K}^{pot}$	Method of determination
CO_3^{2-}	XI	Cl^-	-3.7	SS
		OAc^-	-1.6	
		SO_4^{2-}	-3.8	
		NO_3^-	-0.5	
		ClO_4^-	1.4	
		borate	-1.3	
BF_4^-	V	OH^-	-3.0	MS
		I^-	1.3	
		NO_3^-	-1.0	
		Br^-	-1.4	
		OAc^-	-2.4	
		HCO_3^-	-2.4	
		F^-	-3.0	
		Cl^-	-3.0	
		SO_4^{2-}	-3.0	
NH_4^+	Nonactin–monactin (XXIII) mixture	Li^+	-2.4	
		Na^+	-2.7	
		K^+	-0.9	
		Rb^+	-2.4	
		Cs^+	-2.3	
		Ca^{2+}	-3.8	
		H^+	-1.8	
Li^+	XXIX	Na^+	-1.3	SS
		K^+	-2.2	
		Mg^+	-3.7	
		Ca^{2+}	-3.3	
		NH_4^+	-1.4	
		H^+	-0.1	
Na^+	XXVII	K^+	-1.1	MS
		Mg^+	-0.8	
		Ca^{2+}	-2.2	
		NH_4^+	-2.2	
		H^+	-1.1	
K^+	XXII with XV in dioctylphthalate	Li^+	-4.0	SS
		Na^+	-3.5	
		Rb^+	0.5	
		Cs^+	-0.4	
		Mg^{2+}	-5.1	
		Ca^{2+}	-4.4	
		H^+	-4.4	
K^+	XV	Li^+	-1.0	MS
		Na^+	-1.3	
		Rb^+	0.8	
		Cs^+	0.9	
		Mg^{2+}	-3.1	
		Ca^{2+}	-1.7	
		NH_4^+	-0.4	

Appendix (continued)

Determined	Electrode characteristics	Interferents	$\log k^{pot}_{J,K}$	Method of determination
Ca^{2+}	III	Na^+	-1.7	MS
		K^+	-1.7	
		Mg^{2+}	-1.3	
Ca^{2+}	XVII	Na^+	-5.0	MS
		K^+	-5.2	
		Mg^{2+}	-5.1	
		H^+	0.0	
Ba^{2+}	XX	Li^+	-2.7	
		Na^+	-2.5	
		K^+	-2.0	
		Rb^+	-1.7	
		Cs^+	-1.0	
		Mg^{2+}	-3.6	
		Ca^{2+}	-3.6	
		Sr^{2+}	-2.6	
		Ni^{2+}	-3.9	
		Cu^{2+}	-2.4	
Ba^{2+}	XXI	Ca^{2+}	-3.7	
		Li^+	-3.3	
		NH_4^+	-2.5	
		Cs^+	-2.5	
		Na^+	-2.5	
		Rb^+	-2.0	
		K^+	-2.7	
		Sr^{2+}	-1.5	
		H^+	-1.2	

INDEX